Geography into the Twenty-first Century

Geography into the Twenty-first Century

edited by

Eleanor M. Rawling
University of Oxford

and

Richard A. Daugherty
*University of Wales,
Aberystwyth, UK*

JOHN WILEY & SONS
Chichester · New York · Brisbane · Toronto · Singapore

Copyright © 1996 by John Wiley & Sons Ltd,
Baffins Lane, Chichester,
West Sussex PO19 1UD, England

National   01243 779777
International (+44) 1243 779777

Other Wiley Editorial Offices

John Wiley & Sons, Inc., 605 Third Avenue,
New York, NY 10158-0012, USA

Jacaranda Wiley Ltd, 33 Park Road, Milton,
Queensland 4064, Australia

John Wiley & Sons (Canada) Ltd, 22 Worcester Road,
Rexdale, Ontario M9W 1L1, Canada

John Wiley & Sons (SEA) Pte Ltd, 37 Jalan Pemimpin #05-04,
Block B, Union Industrial Building, Singapore 2057

Library of Congress Cataloging-in-Publication Data

Geography into the twenty-first century / edited by Eleanor M. Rawling
and Richard A. Daugherty.
     p.   cm.
   Includes bibliographical references and index.
   ISBN 0–471–96236–8 (alk. paper)
   1. Geography–Philosophy.   I. Rawling, Eleanor.   II. Daugherty,
Richard (Richard A.)
   G70.G445   1996
   910–dc20
                                                            95–52309
                                                            CIP

British Library Cataloguing in Publication Data

A catalogue record for this book is available from the British Library

ISBN 0 471 96236 8

Typeset in 11/12pt Palatino by Vision Typesetting, Manchester
Printed and bound in Great Britain by Bookcraft (Bath) Ltd, Midsomer Norton
This book is printed on acid-free paper responsibly manufactured from sustainable
forestation, for which at least two trees are planted for each one used for paper
production.

Contents

Preface

This book has its origins in the 1994 Council of British Geography (COBRIG) seminar held in Oxford. In a period of rapid change and increasing pressures on geography and geographical education, the seminar provided an opportunity for reflection, creative thinking and interchange across all levels of education. In this sense it was a new departure and one which amply justifies the existence of COBRIG, a body founded comparatively recently in 1988. The ability of COBRIG to identify the need for such debate, to co-ordinate the wishes of member organisations, and to take the necessary action to make the event happen demonstrates that the Council has a proactive as well as a supportive role to play in relation to the advancement of British geography. The seminar and the book together mark the 'coming of age' of COBRIG and the decision to make the 'COBRIG seminar' a biennial event will strengthen its influence and potential in the coming years.

COBRIG represents all the organisations concerned with geography in England, Wales and Scotland and, increasingly, Northern Ireland is being included in the activities. However, readers of this book will find that the discussions focus mainly on the situation in England and Wales, with briefer reference to Scotland provided mainly through Leslie Hunter's chapter (p. 235). This was necessary in order to make the seminar (and book) manageable, but efforts have been made to draw out issues and concerns which have more general application.

The quality of debate and the overall coherence of the

book are products of the hard work of our colleagues in schools, colleges and institutions of higher education who have written the material despite other heavy commitments. Although based on the seminar presentations, each chapter has been written specifically for this book with the arguments being developed and updated to reflect the changing context of the late 1990s. The Part D chapters, in particular, represent new thinking and dialogue, undertaken since the seminar and stimulated by it.

We have undertaken the editing very much as a joint enterprise, finding this not only an enjoyable experience but one which has helped us to clarify the messages arising from the book. The contributors have thus been subjected to being chased, badgered and advised by not one but two editors! We hope that they did not find the experience too daunting and while thanking them for their efforts, we fully accept responsibility for any errors or inconsistencies.

This book is aimed at a wide readership, including all those interested in the future of geography as a subject as well as those more concerned with it as a medium for education at all levels. Some may wish to delve into individual chapters or to confine themselves to certain parts of the book, whilst others may wish to follow the chapters and the arguments through in a more complete way. In whatever way this book is used, we hope that each reader will find stimulus, enjoyment and, perhaps, new directions for research.

The International Charter, published and disseminated by the Commission on Geographical Education of the International Geographical Union states unequivocally (1993) that 'geographical education is indispensable to the development of responsible and active citizens in the present and future world'. We endorse this statement and believe that future progress will depend on continuing the debate and the action that was begun in July 1994. If this book succeeds in raising important questions and helps to set an agenda for the development of the subject, then we shall be content that it has served its purpose in helping to take geography into the twenty-first century.

Eleanor M. Rawling
Richard A. Daugherty

Acknowledgements

The following sources are gratefully acknowledged.

Figure 11.2   *GNVQ Briefing, June 1995,* National Council for Vocational Qualifications, London.

Figure 11.3   *NCVQ Monitor Summer 1995,* National Council for Vocational Qualifications, London.

Figure 13.1   *A Survey of IT in Geography* (1993), National Council for Educational Technology, Coventry.

Figure 13.2   *A Survey of IT in Geography* (1993), National Council for Educational Technology, Coventry.

Figure 13.3   *A Survey of IT in Geography* (1993), National Council for Educational Technology, Coventry.

Figure 14.1   *Curriculum and Assessment in Scotland: National Guidelines, Environmental Studies,* HMSO, Edinburgh.

List of contributors

Tony Binns   Senior Lecturer in Geography, University of Sussex

Michael Bradford   Senior Lecturer in Geography, University of Manchester

Sue Burkill   Senior Lecturer in Geography, College of St Mark and St John, Plymouth

Graham Butt   Lecturer in Geographical Education, University of Birmingham

Graham Corney   Lecturer in Educational Studies, University of Oxford

Richard Daugherty   Professor of Education, University of Wales, Aberystwyth

John Davidson   Head of Geography, Exeter School

Rita Gardner   Director, Environmental Science Unit, Queen Mary and Westfield College, University of London

Peter Haggett   Professor of Urban and Regional Geography, University of Bristol

David Hall   formerly Lecturer in Geographical Education, University of Bristol

Mick Healey   Professor of Geography, Cheltenham and Gloucester College of Higher Education

Leslie Hunter   formerly HM Inspector of Schools, Scotland

Peter Jackson   Professor of Human Geography, University of Sheffield

Ron Johnston   Professor of Geography, University of Bristol

David Lambert   Senior Lecturer in Education, University of London Institute of Education

Hugh Matthews   Professor of Geography, Nene College, Northampton

Nick Middleton   Lecturer in Physical Geography, University of Oxford

Derek Mottershead   Professor of Geography, Edgehill College of Higher Education

Tim O'Riordan   Professor of Environmental Sciences, University of East Anglia

Eleanor Rawling   Honorary Research Associate, University of Oxford and Professional Officer (Geography) SCAA

Margaret Roberts   Lecturer in Education, University of Sheffield

Tim Unwin   Reader in Geography, Royal Holloway, University of London

Rex Walford   Lecturer in Education (Geography), University of Cambridge

List of abbreviations

Terms

A level	Advanced level of the General Certificate of Education (GCE)
AS level	Advanced Supplementary level of the GCE
AVHRR	advanced very high resolution radiometry
CAL	computer assisted learning
CPVE	Certificate of Pre-vocational Education
CSE	Certificate of Secondary Education
CSYS	Certificate of Sixth Form Studies
GCE	General Certificate of Education
GCSE	General Certificate of Secondary Education
GENE	Geography in Education Network for Empowerment
GENIP	Geographic Education National Implementation Program
GEST	grants for education support and training
GIS	geographical information systems
GNP	gross national product
GNVQ	General National Vocational Qualification
GSIP	Geography Schools and Industry Project
GYSL	Geography for the Young School Leaver
HEI	higher education institution
H grade	Higher grade examination (Scotland)
HIT	Humanities and Information Technology (Project)
INSET	in-service education and training of teachers
IT	information technology
KS	Key Stage (1, 2, 3, 4) of the National Curriculum

LEA   local education authority
NGC   non-geographical competitor
NVQ   National Vocational Qualification
O grade   Ordinary grade examination (Scotland)
O level   Ordinary level of the GCE (England, Wales, Northern Ireland)
PGCE   Postgraduate Certificate in Education
PSE   personal and social education
RAE   research assessment exercise
SCITT   school-centred initial teacher training
S grade   Standard grade examination (Scotland)
TLTP   Teaching and Learning Technology Project
TOMS   total ozone mapping spectrometer
TVEI   Technical and Vocational Education Initiative

Organisations/groups

ABRC   Advisory Board for the Research Councils
ACAC   Awdurdod Cwricwlwm ac Asesu Cymru/ Curriculum and Assessment Authority for Wales (from 1994)
AEB   Associated Examining Board
AGTW   Association of Geography Teachers of Wales
BGRG   British Geomorphological Research Group
BTEC   Business and Technology Education Council
CCEA   Northern Ireland Council for the Curriculum, Examinations and Assessment
CCW   Curriculum Council for Wales (to 1994)
C&G   City and Guilds of London Institute
CERN   European Organisation for Nuclear Research
CHGHEI   Conference of Heads of Geography in Higher Education Institutions
COBRIG   Council of British Geography
DES   Department of Education and Science (to 1992)
DfE   Department for Education (from 1992 to 1995)
DfEE   Department for Education and Employment (from 1995)

EDET	Environment and Development Education Training Group
ESTA	Earth Science Teachers' Association
FEFC(E)	Further Education Funding Council (England)
FEFC(W)	Further Education Funding Council (Wales)
GA	Geographical Association
HEFC(E)	Higher Education Funding Council (England) (from 1992)
HEFC(W)	Higher Education Funding Council (Wales) (from 1992)
HMI	Her Majesty's Inspectorate
IBG	Institute of British Geographers (to 1994)
IEEM	Institute of Ecology and Environmental Management
IUCN	International Union for The Conservation of Nature
IGU	International Geographical Union
IWEM	Chartered Institution of Water and Environmental Management
JET	Joint European Taurus Project
MEG	Midland Examining Group
MSS	Multispectral Scanner
NCC	National Curriculum Council (England) (to 1993)
NCET	National Council for Educational Technology
NCVQ	National Council for Vocational Qualifications
NEAB	Northern Examinations and Assessment Board
NERC	Natural Environment Research Council
OCSEB	Oxford and Cambridge Schools Examination Board
OFSTED	Office for Standards in Education (England)
OHMCI(Wales)	Office of Her Majesty's Chief Inspector (Wales)
RGS	Royal Geographical Society (to 1994)
RGS/IBG	Royal Geographical Society with the Institute of British Geographers (from 1995)
RSA	RSA Examinations Board
RSGS	Royal Scottish Geographical Society

SAGT	Scottish Association of Geography Teachers
SCAA	School Curriculum and Assessment Authority (England) (from 1993)
SCOTVEC	Scottish Vocational Education Council
SEAC	School Examinations and Assessment Authority (England and Wales) (from 1988 to 1993)
SEB	Scottish Examination Board
SEC	Secondary Examinations Council (from 1983 to 1988)
SOED	Scottish Office Education Department
SUCE	Scottish Universities Council on Entrance
TGAT	Task Group on Assessment and Testing
UCLES	University of Cambridge Local Examinations Syndicate
UGC	University Grants Committee (to 1988)
ULEAC	University of London Examinations and Assessment Council
UNCED	United Nations Conference on Environment and Development
UNEP	United Nations Environment Programme
UODLE	University of Oxford Delegacy of Local Examinations
UWCG	University of Wales Council for Geography
WJEC	Welsh Joint Education Committee
WO	Welsh Office
WWF	Worldwide Fund for Nature

Introduction

1 Madingley revisited?

Eleanor M. Rawling

In 1963, the first Madingley Conference, held in Cambridge, provided the creative stimulus which launched an era of change in school geography and which was reflected in *Frontiers in Geographical Teaching* and *Models in Geography* edited by Richard Chorley and Peter Haggett. Thirty-one years later in 1994, the Council of British Geography (COBRIG) hosted a seminar at Mansfield College and the School of Geography, Oxford, once again drawing together academic geographers and schoolteachers to discuss the subject. This book *Geography into the Twenty-first Century* is the result of that dialogue, presenting an edited collection of papers from the seminar.

Delegates at Madingley, Cambridge in 1963 and at Mansfield, Oxford in 1994 shared the same spirit of concern and enthusiasm for the subject. In other respects, however, the two events and the publications which resulted are very different. The Madingley Conference essentially provided the opportunity for geographers from higher education to update their schoolteacher colleagues about new trends and ideas in the subject, particularly about the so-called 'New Geography'. It was an occasion for those in universities to offer 'pearls of wisdom' to those in schools. Although there was plenty of mutual respect, schoolteachers were still viewed as the junior partners in this relationship, not surprisingly since the focus was on research findings and

Geography into the Twenty-first Century, Edited by E.M. Rawling and R.A. Daugherty,
© 1996 John Wiley & Sons Ltd.

new ideas which had been developed in the university sector. The COBRIG seminar had a much wider set of purposes of which updating about the subject was only one. As outlined on the programme for the seminar, the purposes were:

1. to provide the opportunity for geographers in higher education and in schools to share experience and to update each other about new approaches in geography and about significant developments in the school curriculum;

and more specifically:

2. to move towards clarifying the essential character and contribution of geographical education from the secondary years to higher education;
3. to plan future strategies which will ensure continuing dialogue and the well-being of geographical education at all levels.

It will be clear from these purposes that the exchange was very much envisaged as two-way, with updating about school curriculum changes and exchange of views about broader educational objectives being as high on the agenda as developments at the research frontier. The unifying purpose of the event lay in the desire to secure the health of geography at all levels in the education system in a period of rapid change. In this respect, the participants in the seminar were equal partners, whatever their institutional background. The COBRIG seminar 1994 was not a revisiting of Madingley, but a reformulation of the event in a way better suited to the 1990s.

The seminar took place in a context very different from that of the early 1960s. COBRIG is a body which did not exist in 1963 being formed only in 1988, following two decades of increasingly close co-operation between Britain's principal geographical societies. Membership currently includes all those societies and bodies representing geography and geographical education in England, Wales and Scotland (nine bodies in all, see Appendix 1). With its general aim of the advancement of British geography, it seeks to support and complement the work of individual organisations, particularly where matters of policy are concerned or where the subject needs strong representation at national or international level. At the time of COBRIG's formation, for instance, it had been necessary to mount a strong campaign to ensure geography's position in the new National Curriculum in schools. The three specific functions of the Council are, firstly, to provide a forum for discussion and exchange between the member organisations; secondly, to represent the interests of geography and geographers at national and international level; and, thirdly, to identify issues which need investigation or concerted action. The idea for a seminar arose specifically from discussions in Council arising from the third function. COBRIG's action plan for 1993–95 identified a pressing need for schools and higher education to engage in real dialogue about the subject with a view to clarifying its role and character throughout the system. It seemed that the combined pressures of central curriculum change in schools, structural change in higher education and concerns about accountability and resourcing throughout the system were all leading to a situation in which geographers no longer found time to talk about geography. Members of Council were worried about this trend, looking back, perhaps with some nostalgia, to the 1960s and 1970s as a time when there was greater freedom to engage in such a debate. Given the context of the early 1990s, with so many constraints and specific interests relating to each sector of education, it was felt that no one geographical body could easily take the lead on this. It was clearly an opportunity which COBRIG itself should pick up, and so the idea was agreed in the autumn of 1993 and took shape over the subsequent months.

The seminar programme clearly reflected this origin. All member organisations and the main sectors of the education system, from primary teaching to academic research, were represented by appropriate numbers of delegates and as equal partners. Further education was the only sector not well represented by delegates or in discussions, and this was partly because this might have extended the debate too widely for the short time available. However, the further education sector, with its more direct and vocational links to business, industry and the world of work is distinctive, and likely to become increasingly significant now that the Departments of Education and Employment have been merged (July 1995). These matters will be more comprehensively covered in the 1996 COBRIG seminar.

The first part of the seminar programme comprised key

inputs from leading academics and educationalists, although these were short and seen as introductory stimuli to discussion on the Saturday afternoon, not as the 'last word' on the topics concerned. Questions and discussion were taken at the time of the presentations but, more importantly, the whole of the Sunday morning was given over to joint discussion groups. In these, higher education geographers and school geographers exchanged views, debated issues and thought creatively about the next steps, in relation to specific matters which were seen to cross system and institutional boundaries. An important input on the Saturday evening was provided by Professor Peter Haggett, who as a young geographer at Madingley in 1963, had been instrumental in developing that debate. Professor Haggett provided both a link with the past and a stimulating and authoritative view about the future. The final plenary session at the seminar tried to ensure that all delegates went away with a clear view of the further action which was necessary, and which could be initiated either independently or co-operatively by the organisations and individuals present.

The chapters in this book derive directly from that seminar and relate to the programme in the following way. In Part A, 'New perspectives for geography: overview and issues', Professor Peter Haggett's Saturday evening contribution is presented as a thought-provoking link between past, present and future in geography ('Geography into the next century: personal reflections'). This is followed by personal perspectives on 'academic geography' ('Academic geography: the key questions for discussion') from Dr Tim Unwin and on school geography ('School geography: the key questions for discussion') by Dr Tony Binns, who chaired respectively the higher education and the school curriculum presentations on the Saturday afternoon. Part B, 'Geography in higher education' and Part C, 'Geography in the school curriculum' comprise contributions from those who made these short presentations. In each case, the papers provided here are an expanded version of the seminar input, developing the key points made and giving further comment and references as appropriate to the wider readership. Together, Parts B and C chart the state of the discipline as it is taught and studied from schools through to higher education.

In Part B, key developments in human geography are covered by Professors Ron Johnston ('A place in geography') and Peter Jackson, ('Only connect: approaches to human geography'). 'Developments in physical geography' are dealt with by Dr Rita Gardner and environmental issues are considered by Professor Tim O'Riordan in 'Environmentalism and geography: a union still to be consummated'.

In part C, Rex Walford in 'Geography 5–19: retrospect and prospect' looks back to some of the factors influencing the development of the National Curriculum in England and Wales, and forward to his views about the next five years. David Hall's chapter 'Developments at A level' looks more closely at the changing character of geography being promoted through the influential public examination system at 18+. There are many changes currently under way to the whole 14–19 education system and Graham Butt examines these and the implications for geography in 'Developments in geography 14–19: a changing system'. The difficulties inherent in identifying and measuring progression in geography have become more apparent, as teachers have come under increasing pressure to assess and report children's achievements on a national basis. Professor Richard Daugherty's chapter ('Defining and measuring progression in geography') summarises the situation and highlights some important questions for future research and consideration. Sue Burkill considers 'Trends in school geography and information technology' and suggests some implications for higher education, while Leslie Hunter (HMI, Scotland) gives an overview of recent changes and developments in Scotland ('Geography in the Scottish school curriculum'). Finally, Eleanor Rawling reviews 5–19 geography in England and Wales and draws out some key issues for future debate ('Geography 5–19: some issues for debate').

An innovative feature of the book is Part D, 'Learning from the dialogue'. This presents papers which derive from the Sunday morning discussion groups. Each group had a chair and a reporter, one from higher education and one from the school sector. These two individuals have collaborated to write a paper which draws out trends and implications that cross the school/higher education divide. In this respect, this section of the book is unique. Although each individual paper has attempted to summarise the outcomes of the group discussions, the authors have also developed the issues further so that their own joint reflections and interpretations come through. 'Geography at the secondary/higher education interface' by Dr Mike Bradford focuses on the characteristics of the systems and structures which govern the relationship between schools and higher education and so influence the development of geography. Chapters 17, 18 and 19 dealing respectively with 'Human and regional geography in schools and higher education' (Professor Mick Healey and Margaret Roberts), 'The experience of physical geography in schools and higher education' (John Davidson and Professor Derek Mottershead) and 'Teaching environmental issues in schools and higher education' (Graham Corney and Dr Nick Middleton) compare the experience of schools and higher education in developing these aspects and highlight matters for joint debate. Chapter 20 presents an interesting view on 'The contribution of geography to personal and social education' (David Lambert and Professor Hugh Matthews), an area which has been much debated in schools but is only just beginning to surface as a matter for consideration in higher education. If one of the main purposes of the COBRIG seminar was to generate real two-way dialogue, then it is in this section of the book that the dialogue is really seen to come to life. In a positive sense, the issues raised by these authors which have come out of joint discussion, comprise the tentative beginnings of the agenda for future debate and action.

Chapter 21, 'New perspectives for geography: an agenda for action' is an attempt by the two editors (Richard Daugherty and Eleanor Rawling) to draw together the threads from the different sections of the book, to highlight the key issues and to clarify the agenda for the future. It may have seemed to many geographers in schools and in higher education, that the pressures and concerns of the late 1980s and early 1990s would cause a narrowing of perspectives and a stifling of creativity. The 1994 COBRIG seminar and this publication together prove that the debate is alive and expanding. Acknowledgement must be given to all the seminar participants (see Appendix 2) whose enthusiasm and commitment ensured that the debate was well informed and of high quality. The character of geography and its potential contribution to the education of young people of all ages, are at the heart of this book. It is hoped that the reader will be left with an exciting agenda for future action and compelling reasons to continue the dialogue about the future of geography.

Part A

New perspectives for geography: overview and issues

Geography into the next century: personal reflections

Peter Haggett

No one who has had the privilege of taking part in the seminar at Mansfield College can fail to be aware of the debt British geography owes to the Council of British Geography (COBRIG) in general and to Eleanor Rawling in particular. In my own case this debt is shaded by the fact that Eleanor has asked me to deliberate on a topic on which I can be guaranteed to be proved wrong by the many subsequent turns of events. For, at few times has university geography in Britain been changing more rapidly, and in no time has forecasting its future been more difficult.

In this chapter, I should like to sketch out four themes. First, the comparisons with a generation ago. Second, the changes in the university structures within which geography is set. Third, the ways these impinge on a changing geography department. Fourth, the ways in which we may steer a path into the next century.

Some comparisons with Madingley

The Mansfield College summer meeting in Oxford brought back warm recollections of the meetings of geographers at Madingley Hall in Cambridge, some 30 years before. Madingley Hall, an Elizabethan country house near Cambridge, was the scene during the 1960s of a number of

Geography into the Twenty-first Century, Edited by E.M. Rawling and R.A. Daugherty,
© 1996 John Wiley & Sons Ltd.

summer schools for schoolteachers on changes in geography. The summer courses for 1963 were written up as *Frontiers in Geographical Teaching* (Chorley and Haggett, 1965) while subsequent courses led to a second volume entitled *Models in Geography* (Chorley and Haggett, 1967). The idea for the summer schools emerged in 1962, when I was teaching at the University of California at Berkeley and Richard Chorley was at the United States Geological Survey in Colorado; we have recently written about the Madingley experience and I shall not repeat that here (Haggett and Chorley, 1989).

There are several changes which strike me in comparing the two occasions. First, the age of the participants. At the first Madingley conference the average age of the participants (including the two leaders) barely crept into the thirties. This weekend's delegates have more than their fair share of greying hair and balding patches; the average age is probably around 45. If we are still revolutionary, then we are certainly older protesters than those of the 1960s!

Second, British (and indeed, North Atlantic) universities were then in a period of very rapid growth with a progressively greater share of the gross national product (GNP) being channelled into education, especially higher education. To illustrate this growth, I can still recall the dent to my pride when as a University Grants Committee (UGC) member in the 1980s I learnt that the year in which I was appointed to a chair at Bristol (1966) was the *annus mirabilis* of chair appointments. In that particular year far more professorships were created than in any other year since the UGC was founded in 1919. Indeed it was hard not to get a chair that year!

Third, the then predominant positivist paradigm was an optimistic and confident one. Only a decade before, two young scientists at Cambridge—Francis Crick was 36 but Jim Watson only 24—had unravelled the double-helix structure of protein molecules (Watson, 1968). There was a widespread belief abounding in the 1950s that science 'would deliver the goods' and this had spread into geography. Quantitative geography may have had many faults but lack of confidence was then not one of them.

Fourth, space comparisons. The academic landscape of a generation ago was less crowded with scholars, so that there was more space between geographers and their academic neighbours. Today, subjects once taught only within geography schools are now taught in other departments: courses on 'environmental history' are now offered in history departments, 'spatial processes' courses within statistics departments, and so on. The open landscapes of Madingley are now the crowded suburbs of Mansfield.

A fifth difference lay in the relationship between school and university geography. In the Madingley courses, my recollection is that it was the university dons who taught and the schoolteachers who listened (with one or two notable and well-remembered exceptions!). Thirty years on, that hierarchic situation would now be unthinkable. So much information about teaching geography at the school level has accumulated, that the academicians now have much to learn from the schools. Today, it is the schools which are buzzing with ideas and, in any Madingley revival, the situation would be balanced if not reversed.

Taken all round, I sense that the 1990s are a greyer and less certain decade than the 1960s. Economic growth in the university systems of Western Europe and North America has slowed down and there are doubts that scientific technology can deliver a better economic return, at least not without widening the gaps between groups in society (at several different scales). The debates then were about the promise of nuclear power; today the debate has shifted to the problems of nuclear waste.

Changing university structures

I turn now to university structures for these are the matrix within which geography departments are set. We cannot divorce what happens to university geography departments from what happens to the university system as a whole.

Here again, we see a series of strands within the change. First, the social and political pressures for broadening university education. In the UK, a higher proportion of our 18-year-olds are going on to further education, and the need for in-service provision and for third-age needs is growing. The number of fully independent universities in the UK is nearly double the number that it was when Madingley was conceived. This change is still more acute on the world scale, as the elephantine size of reference books like *The World of Learning, 1995* attest. Indonesia, for example, now has some 50 universities.

Second, there are strong economic pressures for university resourcing at no more than a steady state. John Ziman in his

Prometheus Bound (Ziman, 1994) has brilliantly shown how the exponential growth in government-supported research in the 1970s has shifted to an S-shaped curve by the 1990s with every expectation that the rate of growth will progressively slow over the next decade. Ziman argues that this transition is pervasive, interlocking, ubiquitous and permanent. The new policies of 'accountability', 'evaluation', 'priority-setting', 'selectivity', 'critical mass' and the like are not the transient aberrations of a right-wing British government but part of a global response to science's transition to a steady state.

In 1985, Sir David Phillips, then chairman of the Advisory Board for the Research Councils (ABRC), foresaw the British university system splitting into a tripartite division which he termed R, T and X (ABRC, 1987). The first letter stood for research universities, the second teaching universities and the third a mixture of the two. Although the then Secretary of State for Education immediately disowned the RTX concept, I suspect this was because of the wish for a quiet political life rather than on philosophical grounds. Over the last decade very powerful forces have been concentrating research funds into the leading 'R' universities. Research resourcing whether by research councils, government departments or charities is resulting in an ever greater concentration of research in a few institutions.

A fourth trend is towards the movement of research to outside the university system. In some new research areas of relevance to geography, notably geographic information systems (GIS), over 90 per cent of research and development is being conducted outside the university system. Nobel Prizes are no longer confined to academic researchers working within universities; now company-endowed laboratories, like those of Bell Telephones or the Wellcome Foundation, take their share of the world's highest scientific honours.

Even within the state sector, the assumption that learning, teaching and researching form a seamless web is now seriously under review. In Germany over the last few decades, an increasing separation has grown up between the university departments conceived increasingly as teaching institutions, and the Max Planck institutes which are dominantly concerned with research. Some of the research previously conducted by universities is moving into international offshore consortia. In high-energy physics, the

CERN (European Organisation for Nuclear Research) laboratory at Geneva or the JET (Joint European Taurus) laboratories at Culham, Oxfordshire, illustrate such multinational sites. Wherever the heavy capital investment of research outstrips the national capacity to provide it, international consortia are likely to take over.

The changing geography department

Geography departments have not been immune to the changes affecting the university system as a whole. Ziman's move to the steady state and Phillips's RTX model are starting to have serious local implications within our own field. A median-sized geography department in the UK is still only around 15 university teaching officers. If we assume each member has two or three teaching strings to his or her bow (some may be topical, some regional), then this allows not more than between 30 and 45 different teaching areas to be developed in the curriculum.

But on the research side, size has been shown to play an equally critical role as argued by Ron Johnston (1994). If we except one or two lone scholars in each department, then many research projects now have thresholds of two or more members of staff if the group is to be internationally competitive and retain momentum during periods of leave or with changes in personnel. One rule of thumb is that the number of research clusters bears some rough relation to the square root of the staff numbers minus some threshold value. If we set the threshold at two, then it is difficult to maintain more than three research areas with a staff under 15, nor more than five with a staff under 25.

Research advantage further accrues to the larger departments in which leave strategies can be planned, teaching pressures kept under control, and a culture of research built up. Funding bodies are increasingly monitoring and evaluating use of public moneys for research studentships, fellowships, or capital equipment in the search for selectively funded centres of excellence. Staffing in depth and well-funded laboratories are the entry price for many research areas. With too little central research funding to share around, formal recognition is increasingly the gateway for entry into some research areas. Such trends are not just confined to the science side of geography. On the arts side, the old UGC discouraged proliferation in many language areas (Chinese, Japanese, Arabic) because of the need for staff thresholds and the high cost of small specialised language libraries (Parker, 1986). The Higher Education Funding Council (HEFC)'s mechanism for allocating resources may well exacerbate these tendencies (Johnston, 1995), so that in future there will be fewer research centres operating at ever lower unit costs.

A second fact of life for geography departments is the growth of non-geographical competitors (NGCs). These include subjects such as environmental history (sponsored by history departments), environmental appraisal (deriving from planning and management) and spatial analysis (developed from operations research). It was said of the Berkeley geographer, Carl Sauer, that he was like a voortrekker farmer in moving on his research camp if he saw too many twinkling lights from other researchers in the distance. Today the lights are now twinkling all round the geographical camp in the earth sciences, in law, in economics, in statistics. How to live with the NGCs is an important policy problem. Lofty separation leads to duplication of effort and loss of opportunity. Co-operation is usually the preferred policy but it demands that the nature of our geographical contribution is made clear.

Steering a course into the next century

In the Bristol seminars which led to his introduction to *Explanation in Geography* (Harvey, 1969), Harvey drew an analogy between a discipline and a ship. A department needs to be large enough to be seaworthy, have an engine powerful enough to move forward, have a sensitive steering mechanism to set a course, and have an agreed course by which to steer. In this comparison, the funding council or the Department for Education (DfE) can be viewed as generating cross-winds and tides: in good time sweeping us forward, but in bad times serving as head winds or cross-buffeting.

In that setting, the critical question relates to the course we wish to steer. With Rex Walford, I have set out recently some of the directions in which geography might be steering in the next few decades and I do not wish to repeat them here (Walford and Haggett, 1995). But the concern with global issues, the use of information technology (IT), the redesigning of regional geography and the interlinking with philosophical and ethical issues all remain critically important areas to which geography needs to contribute.

But universities are not just about advancing knowledge, they are also concerned with conserving knowledge. In the face of increasing change we need to conserve and re-emphasise some central and cherished aspects of geographical education; a love of landscape and of field exploration, a fascination with place, a wish to solve the spatial conundrums posed by spatial configurations. We need to reinforce the existing links between geographical study at different levels and to make more transparent the interfaces between school, college, university and research institutes.

Let me close on the last point. I have already indicated that the balance between school and university has already changed and may change further. We need only look west across the Atlantic, to see how fortunate we are here in the UK to have a lively and scholarly school sector to complement and underpin the university curriculum. But there remains the need for further integration. Universities need to be ever sensitive to the many changes in the school curriculum. We are already feeling the ill effects of a reduced competence in physical geography in the school curriculum. Equally we benefit by the many strides being made in computer competency. If we are to get the best of both systems we need to agree about the geographic fundamentals at both school and university level. We also need to see how both sectors can fit into the burgeoning private sector in geography and in this respect we need to be very sensitive about how the several pieces fit together. COBRIG is one of the few organisations well placed to fill this gap and I wish it well.

References

ABRC (Advisory Board for the Research Councils) (1987) *A Strategy for the Science Base*. HMSO, London.

Chorley, R.J. and Haggett, P. (eds) (1965) *Frontiers in Geographical Teaching*. Methuen, London.

Chorley, R.J. and Haggett, P. (eds) (1967) *Models in Geography: the Second Madingley Lectures*. Methuen, London.

Haggett, P. and Chorley, R.J. (1989) From Madingley to Oxford: a foreword to *Remodelling Geography*. In W. Macmillan (ed.), *Remodelling Geography*. Blackwell, Oxford, pp. xv–xx.

Haggett, P. (1990) *The Geographer's Art*. Basil Blackwell, London.

Harvey, D. (1969) *Explanation in Geography*. Edward Arnold, London.

Johnston, R.J. (1994) Department size, institutional culture and research grade. *Area*, **26**, 343–350.

Johnston, R.J. (1995) Nasty, brutish and short, in Opinion, *Times Higher Education*, 9 June, Times Newspapers Ltd, London.

Parker, P. (1986) *Speaking for the Future: a Review of the Requirements of Diplomacy and Commerce for Asian and African and Area Studies*. UGC, London.

Walford, R. and Haggett, P. (1995) Geography and geographical education: some speculations for the twenty-first century. *Geography*, **80**(346), 3–13.

Watson, J.D. (1968) *The Double Helix*. Weidenfeld & Nicolson, London.

Ziman, J. (1994) *Prometheus Bound: Science in a Dynamic, Steady State*. Cambridge University Press, Cambridge.

Academic geography: the key questions for discussion

3

Tim Unwin

Academic geography: a question of meaning

The chapters in Part B of this book provide a comprehensive overview of the contemporary practice of geography in institutions of higher education and research institutes. This introductory chapter on 'academic geography' seeks to take a different perspective and to explore some of the implications for geography of broader changes taking place in British society. In particular, it addresses recent changes that have occurred in educational practice and legislation. In so doing, it focuses on three main issues: the effects of curriculum changes in primary and secondary geography on the practice of university education; the potential and problems of greater linkages between geographers and those in commerce and industry; and the implications of increased quality control practices in higher education institutions on teaching and research in geography. The overall theme of the chapter is that it is important for geographers in higher education to become increasingly active in the social, economic and political environment within which they work if they are to maintain control over their teaching and research.

First, though, it is essential to explore the meaning of the title chosen by the editors for this particular chapter. My first reaction, on being asked to contribute to this volume, was

Geography into the Twenty-first Century, Edited by E.M. Rawling and R.A. Daugherty,
© 1996 John Wiley & Sons Ltd.

that there was no such thing as 'academic geography' and, therefore, that it was going to be far from easy to embark on such a chapter (although see Johnston, 1986). But naming something gives it reality; a name literally gives meaning to an object, and separates it from other things identified by other names. The problem is, as Gunnar Olsson (1991, p. 15) has pointed out, that 'to define is to take the first step towards the fallacy of misplaced concreteness and the alienation it embodies'. Thus, by calling something 'academic geography', one in a sense brings it into being, but at the same time one tends also to give it a concrete reality which can be misleading. Many would thus argue that there is no such single thing, nor even has been, as geography. Indeed, the title of Ed Soja's (1989) *Postmodern Geographies* accords specific recognition to the idea that many geograph*ies* exist. Moreover, to call something 'academic geography' separates it from other kinds of geography. Within this book, two broad divisions are thus postulated, between geography *in the school curriculum*, which is seen as being an *educational* perspective, and geography in *higher education*, which is considered as being the *academic* or *research* perspective. This would seem to imply that geographers in higher education institutions are not particularly interested in education, but rather concentrate on academic or research-oriented agendas. There is much evidence to support such a viewpoint. Thus Ron Johnston's (1993a) recently edited book, entitled *The Challenge for Geography*, which specifically deals with the question of how geographers should respond to a fast-changing world, pays very little attention at all to educational issues, or to the practice of geography in schools. In contrast, this chapter seeks to explore the interfaces between such differences, and thus the connections between both secondary and higher education geography, and also between higher education and the broader society of which it is a part.

The use of the word 'academic' is also particularly interesting in another way. In origin, the word was applied to the school of philosophy of Plato (428/7–348/7 BC), and subsequently came to be used to refer to that which pertained to a learned society or university. However, it can also have the specific meaning of something that is not practical, or fails to lead to a decision. Taken together these meanings have often been seen to imply that much of what happens at universities is irrelevant to daily life; the image of universities as 'ivory towers' is still one that is prevalent, despite the dramatic changes that have taken place in British higher education institutions over the last 20 years. This chapter challenges such a view, and argues instead that the practice of geography is deeply embedded in the social, economic and political realities of contemporary life, and that geographers have a very significant role to play in changing these realities (Unwin, 1992).

The secondary–tertiary interface

Andrew Goudie, as President of the Geographical Association (GA) in 1993–94, bemoaned the gulf that he saw as having emerged between schools and universities. He thus argued that 'A chasm has developed between those who teach at school and those who teach in universities' (Goudie, 1993, p. 338). He went on to identify three possible reasons for the widening gulf: an increasing desire among academics to do research, a burgeoning of textbooks, and changes in the GA itself. While Goudie appears to conflate membership of the GA with interest and involvement by university geographers in schools, there have undoubtedly been changes in the relationships between these two sectors of geographical education.

In introducing the COBRIG July 1994 seminar of which this book is one outcome, Eleanor Rawling drew attention to the powerful influence of the 1963 Madingley lectures organised by Peter Haggett and Richard Chorley. Fired by what they had seen as 'exciting new developments in the universities on both sides of the Atlantic', Haggett and Chorley (1989, p. xv) had been eager to pass these 'on to young geographers struggling through their sixth-form courses'. Both the lectures and the resulting book *Frontiers in Geographical Teaching* (Chorley and Haggett, 1965) were to have a lasting impression on a generation of geography teachers. Chorley and Haggett's (1965, p. 377) emphasis on what they termed 'the neglected geometrical side of the discipline' was to play a major role in ensuring that a quantitative approach, in which models and theories had a significant emphasis, filtered down into school textbooks and examination syllabuses. Moreover, many schoolteachers entering the profession at that time had been educated in universities where these views were becoming increasingly accepted as the most promising way forward for the discipline.

While it is difficult to determine the full extent of this influence across Britain as a whole, and also its duration, the continued publication of theory- and model-based texts, such as Bradford and Kent's (1977) *Human Geography: Theories and Applications*, into the late 1970s and early 1980s suggests that the teaching of secondary geography in the latter part of the 1960s and throughout the 1970s and early 1980s remained heavily influenced by the research agendas of university geographers.

By the end of the 1970s, though, a dramatic change was occurring. Secondary education was increasingly being seen by society in general, and by the teaching profession in particular, as being concerned with very much more than simply a training for university entrance. Advances in educational theory and practice, including greater attention to curriculum design, assessment methods and teaching skills, meant that increasing emphasis was being placed on the needs of the vast majority of pupils who were never to go on to higher education. Consequently, the practice of teaching within schools, and the learning environments of pupils within the primary and secondary educational systems, were becoming increasingly separated from the practice of university research and teaching.

Within geography, one of the clearest and most important expressions of this change was the development of the Geography 16–19 Project which began at the University of London Institute of Education in 1976 (Naish and Rawling, 1990). This provided curriculum developers with a clean sheet, enabling them 'to ask basic questions about the needs of students and about how geography could contribute towards fulfilling those needs. Development work could then take place to try to construct geography courses pertinent to the students taking them, and to the world in which those students and their teachers live' (Naish and Rawling, 1990, p. 58).

The resultant innovative and enquiry-based GCE A level course was first examined in 1982, and by 1991 over 10 000 candidates sat the examination. A direct effect of this for teachers in higher education was that by the late 1980s growing numbers of incoming students were coming from this new A level, with very different skills and knowledge backgrounds from those of their traditional A level entrants. Some university geographers bemoaned what they saw as a diminution in the knowledge levels of these new first-year undergraduates, while others blamed the Geography 16–19 Project for a demise in the scientific skills necessary for undergraduates satisfactorily to undertake university courses in physical geography. The most severe critics condemned the 16–19 Project as 'laying the subject open to criticisms of superficiality from the natural scientists' (Bailey, 1989, p. 156). Such developments, though, were among the first to provide warnings, for those in institutions of higher education alert enough to see them, that substantial changes would soon have to be wrought in their long-established practices.

Another good indication of the demise of university influence on primary and secondary curriculum design, is the lack of participation by university geographers in the introduction of the National Curriculum in England and Wales. A low level of academic representation was apparent across all subjects, but while some geographers in institutions of higher education contributed to the process through their representation on working groups of bodies such as the GA and the Institute of British Geographers (IBG), there was remarkably little active involvement by them in the actual development of the geography National Curriculum. In large part, this was a result of the way in which the government set up the Working Groups. Thus the composition of the 12-member Geography Subject Working Group, given the task of specifying the precise attainment targets and programmes of study for geography in the National Curriculum, included only one practising academic in a department of geography in an institution of higher education. Other members were a regional secretary of the Country Landowners' Association, the chairman of a travel company, and a member of the Countryside Commission. It is difficult to envisage in retrospect what effect there might have been if there had been greater higher education involvement in the development of the National Curriculum, but it is, at least, possible that there might have been greater emphasis on some of the more exciting and innovative work that has recently taken place in human and physical geography, similar to that highlighted in the chapters of Part B of this book.

From a position of considerable influence on the school curriculum in the 1960s, those in university departments of geography now have little effect on what is taught at the secondary or primary level. While some university geographers continue to play a role in A level examining, it is worrying how few are aware in any detail of the enormous changes that have taken place in the secondary education system in recent years. In part this is a result of their more pressing need to come to terms with the changes that have taken place in the higher education sector, and which are discussed in the final section of this chapter. However, it is essential that good communications are established between those in the secondary and tertiary sectors so that the most appropriate learning experiences can be provided for the growing number of students passing from the one to the other (Unwin, 1989). While the geography that is taught at the primary and secondary level should provide one of the essential elements of an education for life (Bailey and Binns, 1987), it should also provide a sufficient grounding for those wishing to embark on a geography degree to be able to achieve the highest levels of analysis and research.

The lack of knowledge by university geographers about the educational backgrounds of incoming undergraduates is a problem of sufficient magnitude by itself. However, with the rapid increases in the numbers of geography undergraduates since the mid-1980s, this problem has been greatly exacerbated. Moreover, increases in student numbers have not been matched by additional resources and commensurate further staff appointments. Jenkins and Smith (1993) thus report that student–staff ratios in British higher education geography departments increased from an average 12.1 : 1 in 1986 to 17.4 : 1 in 1991. Some of the conclusions of their survey of the effects of this increase on teaching quality are worth quoting at length (Jenkins and Smith, 1993, pp. 508, 509):

> Many staff perceived increased sizes of classes/more students as providing significant problems for students. Seminars, laboratories and practical, dissertation supervision and fieldwork were seen as being adversely affected.
> There was a strong view that staff could give less time to individual students with academic or personal problems.

In summary, they note 'the overwhelming conclusion from both the statistical and qualitative returns is that most aspects of the teaching in geography departments have been adversely affected by having to teach more students and larger classes' (Jenkins and Smith, 1993, p. 509).

While staff believe that increasing numbers have adversely

affected students' learning experiences, this is not reflected in the quality of degree results gained by those students. Indeed, across the discipline as a whole there has actually been a trend towards substantially higher percentages of upper seconds (Chapman, 1993). Jenkins and Smith (1993) interpret this as an indication of the way in which staff have coped with these changes, through the introduction of innovative teaching methods designed to maintain quality (see also Johnston, 1994). One other factor influencing the apparent increase in the number of better quality degree results to which they do not allude, though, has been the manipulation of the systems by which such classifications are awarded. Although no overall survey of geography degree allocation systems has yet been undertaken, there is anecdotal evidence to suggest that in some institutions the systems of awarding particular degree classifications have been adjusted in order to improve the results. When universities are increasingly being judged by quantitative criteria, including variables such as the percentage of good degrees (firsts and upper seconds) that they award, it is scarcely surprising that such actions are being taken. Although such statistical quirks mean that it is extremely difficult to objectify standards, there are nevertheless indications that some geographers have been able to develop new ways of teaching larger numbers of students, while maintaining or even improving their learning experience. Gibbs and Jenkins (1992) have thus produced a major edited volume, building on the experiences of geographers across the higher education sector, on ways of teaching larger classes without any diminution in the quality of the learning experience. In general, despite such signs of innovation, there is little evidence that the majority of teachers of geography in British universities have yet adapted to the government's desired implementation of a 'mass higher education system' (Jenkins and Smith, 1993, p. 513). Most are still trying to adapt their old teaching methods to the new circumstances, rather than implementing wholesale structural changes to their degree programmes.

Several interrelated changes suggest that there is a need to redefine the interface between schools and higher education, and for those in both secondary and higher education to rethink the nature of degree courses and the way in which they are taught. The introduction of a National Curriculum for geography pupils aged 5–14, the development of new

GCE A level syllabuses, and an aspiration for there to be a greatly increased proportion of young people entering higher education, all point to the need for there to be a substantial revision of undergraduate geography teaching programmes. It may well be that this will lead to the introduction of North American teaching methods and styles of assessment, associated with the expansion of Masters' programmes for the more able students. Signs of this are already beginning to emerge, but one key question for debate over the coming years is likely to be how we can effectively enhance the quality of learning in a geography degree for the much broader character and increased size of the student intake, while maintaining levels of research excellence.

Applied research and commercial linkages

One result of the increased financial pressure under which most university institutions have found themselves since 1980, has been that they have been forced to turn increasingly to non-traditional sources for funding. Most often this has been through contract research and consultancy, but opportunities also exist for the commercial funding of teaching (Unwin, 1995). As department budgets become increasingly restricted, as competition for limited numbers of Research Council grants becomes ever more intense, and as the effective value of student grants is further eroded, geographers are increasingly having to turn to external sources of funding for their research and teaching. Traditionally, geographers have had many fewer linkages with commerce and industry than have, for example, engineers, chemists and economists, but there are great advantages to be gained from a closer *rapprochement*. Ron Cooke's (1992; p. 149) warning needs to be heeded: 'In this brave new world, financial competition in the market place is likely to diminish, not enhance, academic collaboration: there is less room for altruism, and it is the best entrepreneurs, rather than the best scholars, who are the most likely to succeed. We shall have to fight if we wish to retain any independence.'

As pointed out in the introduction to this chapter, the very word 'academic' is frequently used to refer to things that are not practical. By definition, therefore, at the heart of our usage of the word in its association with universities, there is an underlying conceptualisation that the 'academic' is separate from the daily life of the 'applied'. Traditionally, 'pure' research has thus been distinguished from the 'applied', and has generally also been accorded more status. In part this is related to deep-seated social conditioning, and parallels the hierarchical distinction between different forms of labour, in which, for example, mental activity is usually accorded greater status than manual labour. Nevertheless, there is a healthy tradition of applied research within geography, and particularly within geomorphology. Gregory (1985) has thus drawn attention to the increasing emphasis on questions of an applied nature in physical geography, a trend which Cooke (1987) sees as being a result of four interrelated factors: a shift in research emphasis towards contemporary processes, an improvement in techniques enabling appropriate advice to be given to clients, the increasing international concern with environmental issues, and a desire by growing numbers of geomorphologists to serve the wishes of those managing the environment. Likewise, the explosion of commercial interest in spatially referenced data and geographical information systems (GIS) reflects another area where geographers have considerable applied experience (Rhind, 1993).

An increasing emphasis on applied research is often seen as a necessary, but unwelcome trend. In particular, it is widely held that excessive control over research by external agencies can be detrimental to originality and academic freedom; in short, that it restricts critical enquiry (Unwin, 1992). In contrast to such views, it can be argued that there are distinct benefits to be gained from closer linkages between geographers and others in the community of which they are a part. Four illustrations can be given of ways in which such linkages can be mutually beneficial in terms of teaching and research.

First, on very practical grounds, there are sound reasons why geographers should seek closer links with the employers of their graduates. It is a frequent complaint of employers that graduates lack many of the skills that they require in their subsequent careers (Roizen and Jepson, 1985; Unwin, 1986), and employers thus have much to gain from the enhancement of relevant training as part of an undergraduate degree. Moreover, geographers can learn much from the training programmes provided by some employers. From a pragmatic viewpoint, departments that have a proven track record of placing their graduates in employment are also likely to benefit from higher levels of applications, as potential undergraduates look ever more closely at the relative benefits of studying for degrees in different departments. As Healey (1992, p. 7) has pointed out, geographers have been at the forefront of developments designed to promote enterprise in higher education, in particular through 'the incorporation of personal transferable skills into degree courses', but much still remains to be done in terms of bringing the teaching requirements of industry and geographers closer together. There are as yet few geography courses which include an industrial sandwich element to them (although see Clark et al., 1990), or even which specifically offer vacation employment opportunities for students in industrially or commercially related fields (but see Unwin, 1995). Furthermore, the development of such teaching linkages can frequently lead to possibilities for research collaboration, opening up opportunities which might not otherwise have existed.

Second, the direct funding of research by industry can have valuable financial benefits for both individuals and departments. In particular, the profits from such research can be used to fund other research activities for which no other source of support might be available. Traditionally, there has been a tendency in universities to undercharge for consultancy and research activities, and many companies have benefited from this as a cheap way of gaining research expertise. Moreover, the financial undervaluing of university research may well have been one factor leading to the low opinion of academic research in general, held by many industrialists. Geographers are well placed to undertake a very wide range of research consultancy, but in order to gain such contracts we need to be much more aggressive in the propagation of our skills and abilities. We have in the past possibly been over-concerned about justifying the existence of geography as a field of enquiry, and have failed to see sufficiently clearly the potential contributions that we can make to social and economic change. If we do not believe in ourselves, then it is hardly surprising that those in the wider community will also have doubts!

Third, there are many possibilities for geographers to obtain direct sponsorship for their research and teaching, often with few, if any, strings attached. The issue of patronage and sponsorship is nevertheless a highly contentious one, with views on it ranging from those who feel that any such patronage is a corrupting influence reducing academic freedom, to those who are willing to take any kind of financial support to enhance their activities. There are fundamental ethical and ideological issues involved here. Most recently and visibly, these have been reflected in the debate over corporate patronage following the merger of the Institute of British Geographers with the Royal Geographical Society. The Royal Geographical Society prior to the merger had been successful in attracting substantial support from Land Rover, British Airways and Shell as corporate patrons, but with the execution of Ken Saro-Wiwa by the Nigerian government in November 1995, Shell's corporate patronage of the joint Society was vociferously condemned by a significant number of former members of the Institute of British Geographers. There, nevertheless, remain many other instances where companies or charities have been willing to provide small amounts of direct sponsorship for use by geographers. This has been, perhaps, easiest with respect to expeditions, but close targeting and persistent effort can also be rewarded in other areas. With the tightening of departmental budgets for undergraduate field courses (Gray, 1993), for example, it is often possible to benefit from sponsorship arrangements in order to defray some of these costs (Unwin, 1995).

Fourth, though, it is important to return to a more theoretical level, and explore some of the justifications upon which academics have traditionally based their privileged rights to so-called 'academic freedom' and independence. This is not the place for a detailed discussion of these issues, but it is likely that in the coming years such questions are increasingly going to be brought to the fore as university academics have to defend their status and positions. The importance of the pursuit of knowledge for its own sake is a concept that is regularly being challenged, for example, as higher education institutions are being encouraged by government to focus more attention on the provision of technical skills that are deemed to be of relevance to society in general (Unwin, 1992). Likewise, while 'pure' research may well have practical applications in 50 or 100 years' time, the high cost of much of this work in the physical sciences means that at times of economic stringency it is often difficult to attract funding for it unless there are clear indications of some short-term benefits. Furthermore, and to put it somewhat crudely, we need constantly to recall that most academic geographers' salaries are paid for largely from taxpayers' pockets. Put in this light, our responsibility is above all to them, and a strong case can be made for geographers to become involved much more closely in local research projects, whose agendas and objectives are determined through a dialogue between communities and the academics who live and work in them.

The central point that needs to be made is that academics have always acted within a set of constraints, although the character of these constraints has varied over time. Government-funded 'pure' research, however this is defined, must remain a central component of individuals' and departments' research income strategy. However, if we seek to undertake research on subjects for which it is difficult to obtain funding from government or Research Councils, then we need to explore other avenues of support. Above all we need to be innovative and creative in our research and teaching, not expending too much energy in bemoaning externally imposed changes, but rather seeking to create our own freedom of choice and independence.

Quantifying performance: the bureaucratisation of higher education

One of the most striking changes to have influenced higher education institutions in recent years has been the much greater accountability imposed on them by government. Thus, one of the factors behind Goudie's (1993) comment that academic geographers are increasingly turning to research rather than involvement with schools, is the importance placed by universities on the research selectivity exercise, first undertaken in 1985. Universities and disciplines are now subject to three types of quality control: audit, through which an institution's quality control mechanisms are examined; research assessment (RAE), through which departments' research quality is assessed; and quality assessment of the education provided by higher and further education institutions. These have had far-reaching effects on the institutional practice of geography, and have evoked strong feelings both of support and of criticism.

What is undeniable about the introduction of such quality control mechanisms is that they have taken up a considerable amount of staff time, which had previously been available for research and teaching. The completion of self-assessment forms, the submission of reports for the research selectivity exercise, and the time actually spent by those undertaking the assessments, all bite deeply into the working lives of university geographers. The critical question that therefore needs addressing is whether or not they have actually improved the quality of research and teaching undertaken; there can be little doubt that they have reduced the quantity of such output for those involved closely in the assessment process, and that they have also been a drain on financial resources which could have been allocated elsewhere.

Of the three types of quality control, the research assessment, or research selectivity, exercise was the first to be introduced, and is the one that has been most discussed. So far there have been three such assessments for geography, in 1985, 1989 and 1992, although the rules by which they have been undertaken have changed each time. For those unfamiliar with it, the importance of the exercise can be put quite simply: the grade allocated to a department determines the level of funding received by an institution. As Ron Johnston (1993b, p. 179) has commented, 'The implications of the financial and other decisions involved in participation in the RAEs are manifold, not only for individual departments and institutions, but also for the discipline and the nature of higher education itself. "Publish or perish" now has a £ sign alongside it to add to the threat . . . at least we know that, even if the detailed parameters are altered, the assessed quality of our research matters—to the tune of up to several tens, if not hundreds, of thousands of pounds for each department.'

It is, nevertheless, far from easy to determine the effects that these three research assessment exercises have had on the practice of research by geographers. Keith Clayton (1993), for example, is in no doubt that they have done little to persuade geographers to enhance their research. In 1985 he undertook a survey of heads of geography departments and 'found little evidence . . . that more than a handful had any serious interest in research, or intended to try and improve various measures of their research performance' (Clayton, 1993, p. 189). Following the latest research assessment exercise, he commented as follows: 'We have now had two research assessment exercises and the latest shows that geography is not in a particularly strong position. I suspect, in line with my earlier survey, that few outside the top-grade departments are much troubled by this result. My view is reinforced by evidence that several departments have not grasped recent opportunities to improve their research grade—indeed, they may deliberately have chosen the traditional geographical route of reaffirming their dedication to teaching and their defensive belief that improved research performance can only be at the expense of teaching quality. That is, of course, utter nonsense' (Clayton, 1993, p. 189). This is an extreme view, but it nevertheless sheds doubt on the overall efficacy of the research assessment system in improving geographical research quality, however that is defined. What the exercise has certainly done in most, if not all, departments, is to improve the information systems about their research activities, measured in terms of quantifiable criteria such as grant income and number of publications. The emphasis of the review panel in the 1992 exercise on publications in high-quality international journals (Gregory, 1993) has also focused the efforts of some geographers on getting their research published in these particular sources of output.

The early emphasis on research assessment drew attention to a division between research and teaching. Colin Thorne (1993, p. 169) has thus argued that 'The result of the funding system is that, depending on the present and future strategy and research performance of a lecturer's institution and department, he or she will very soon either be exploring the sunlit research uplands, or toiling deep in the teaching salt mines. It is *that* important. Of course, both are perfectly worthy occupations, but they are worlds apart and this needs to be recognised now.' The introduction of quality assessment of education may well redress this balance somewhat. The results of the assessments for geography in Scotland were published in the summer of 1994, and those for England and Wales undertaken between autumn 1994 and spring 1995. The effects of this exercise on the quality of learning experience provided for geography students will not be clear for a number of years, but it is already apparent in the completion of self-assessments by departments and in the preparation of supporting documentation, that geographers are being forced to look more closely at their teaching provision. Many would say that this attention is long overdue, but this too is probably an overstatement of the situation. As pointed out above, geographers have been at the forefront of many educational innovations (Healey, 1992), and have a long tradition of concern with education, reflected for example in the activities of the Higher Education Study Group of the Institute of British Geographers, the work of the editorial board of the *Journal of Geography in Higher Education* (Gerber, 1992), and the publication of major volumes on teaching in higher education (Gold et al., 1991). It is to be hoped that one effect of quality assessment will be to share such good practice more widely among the community of geographers in higher education.

Whatever the debates over the difficulty of quantifying the quality of research and teaching, it is evident that, at least in the short term, these are likely to remain important aspects of life for geographers in universities. There is no denying that users of public funds, such as universities, should be accountable for the ways in which they are spent, but there is considerable debate as to whether or not the systems as they currently exist are achieving the aims for which they were intended. Moreover, there is a real danger that they will constrain the diversity of practice of research and teaching, as universities, departments and individuals increasingly seek to satisfy the parameters of excellence as defined by those responsible for the assessments.

Conclusion

This chapter has presented a personal view of three of the key issues currently influencing the practice of geography in British higher education, focusing in particular on the interface between schools and universities. The original purposes of the COBRIG seminar were to clarify the character and contribution of geographical education from the secondary years to higher education, and to plan future strategies to ensure a continuing dialogue and the well-being of geographical education at all levels. Set against this context, the key conclusions of this chapter are twofold: that so-called 'academic' geography should not be seen as isolated from the everyday world of lived experience, but instead should be advocated as of real practical value and relevance; and secondly that 'academic geography' is just as much about education and learning as is school geography. As Goudie (1993) has argued, it is essential for teachers in higher education to maintain a positive engagement with their colleagues in the secondary sector, but at the same time they also need to create the context within which the research priorities in which they believe can be advanced. The world of higher education is not simply *academic*; much geographical research is of profound contemporary importance. We need to convince others that this is so, both through our everyday research and teaching practice, and also through our involvement in the wider world of industry and political action.

Acknowledgements

I would like to take this opportunity gratefully to thank Ron Johnston, Paul Clark, Eleanor Rawling and Richard Daughtery for their most useful comments on an earlier draft of this chapter.

References

Bailey, P. (1989) A place in the sun: the role of the Geographical Association in establishing geography in the National Curriculum of England and Wales, 1975–89. *Journal of Geography in Higher Education*, **13**(2), 149–157.

Bailey, P. and Binns, T. (eds) (1987) *A Case for Geography: a Response to the Secretary of State for Education from Members of the Geographical Association*, the Geographical Association, Sheffield.

Bradford, M.G. and Kent, W.A. (1977) *Human Geography: Theories and Applications*, Oxford University Press, Oxford.

Chapman, K. (1993) Degree results in geography 1973–1990: students, teachers and standards. *Area*, **25**(2), 117–126.

Chorley, R.J. and Haggett, P. (eds) (1965) *Frontiers in Geographical Teaching*. Methuen, London.

Clark, D., Healey, M.J. and Kennedy, R. (1990) Careers for geographers. *Journal of Geography in Higher Education*, **14**, 137–149.

Clayton, K. (1993) Lack of ambition revisited. *Journal of Geography in Higher Education*, **17**(2), 189–191.

Cooke, R.U. (1987) Geomorphology and environmental management. In M.J. Clarke, K.J. Gregory and A.M. Gurnell (eds) *Horizons in Physical Geography*. Macmillan, Basingstoke, pp. 270–287.

Cooke, R.J. (1992) Common ground, shared inheritance: research imperatives for environmental geography. *Transactions of The Institute of British Geographers*, New Series **17**(2), 131–151.

Gerber, R. (1992) Is JGHE a quality journal? *Journal of Geography in*

Higher Education, **16**(2), 235–237.

Gibbs, G. and Jenkins, A. (eds) (1992) *Teaching Large Classes: Maintaining Quality with Reduced Resources*. Kogan Page, London.

Gold, J.R., Jenkins, A., Lee, R., Monk, J., Riley, J., Shepherd, I, and Unwin, D. (1991) *Teaching Geography in Higher Education: a Manual of Good Practice*. Basil Blackwell, Oxford (Institute of British Geographers Special Publications Series, 24).

Goudie, A. (1993) Guest editorial: schools and universities—the great divide. *Geography*, **78**(4), 338–339.

Gray, M. (1993) A survey of geography fieldwork funding in the 'old' UK universities, 1990–91. *Journal of Geography in Higher Education*, **17**(1), 33–34.

Gregory, K.J. (1985) *The Nature of Physical Geography*. Edward Arnold, London.

Gregory, K.J. (1993) The 1992 UFC research assessment exercise for geography. *Journal of Geography in Higher Education*, **17**(2), 170–174.

Haggett, P. and Chorley, R.J. (1989) From Madingley to Oxford: a foreword to *Remodelling Geography*. In B. Macmillan (ed.) *Remodelling Geography*, Basil Blackwell, Oxford, pp. xv–xx.

Healey, M. (1992) Curriculum development and 'enterprise': group work resource-based learning and the incorporation of transferable skills into a first year practical course. *Journal of Geography in Higher Education*, **16**(1), 7–19.

Jenkins, A. and Smith, P. (1993) Expansion, efficiency and teaching quality: the experience of British geography departments. *Transactions of the Institute of British Geographers*, new series **18**(4), 500–515.

Johnston, R.J. (1986) Four fixations and the quest for unity in geography. *Transactions, Institute of British Geographers*, new series **11**(4), 449–453.

Johnston, R.J. (ed.) (1993a) *The Challenge for Geography. A Changing World: a Changing Discipline*. Blackwell, Oxford (Institute of British Geographers Special Publications Series, 28).

Johnston, R.J. (1993b) Removing the blindfold after the game is over: the financial outcomes of the 1992 research assessment exercise. *Journal of Geography in Higher Education*, **17**(2), 174–180.

Johnston, R.J. (1994) Resources, student: staff ratios and teaching quality in British higher education: some speculations aroused by Jenkins and Smith. *Transactions, Institute of British Geographers*, new series **19**(3), 359–365.

Naish, M. and Rawling, E. (1990) Geography 16–19: some implications for higher education. *Journal of Geography in Higher Education*, **14**(1), 55–75.

Olsson, G. (1991) *Lines of Power/Limits of Language*. University of Minnesota Press, Minneapolis.

Rhind, D. (1993) Maps, information and geography: a new relationship. *Geography*, **78**(2), 150–160.

Roizen, J. and Jepson, M. (1985) *Degrees for Jobs: Employers' Expectations of Higher Education*. Society for Research into Higher Education and NFER-Nelson, Guildford.

Soja, E. (1989) *Postmodern Geographies: the Reassertion of Space in Critical Social Theory*. Verso, London.

Thorne, C. (1993) University Funding Council research selectivity exercise, 1992: implications for higher education in geography. *Journal of Geography in Higher Education*, **17**(2), 167–170.

Unwin, T. (1986) Attitudes towards geographers in the graduate labour market. *Journal of Geography in Higher Education*, **10**(2), 149–158.

Unwin, T. (1989) From secondary to higher education: overlap or divide? *Area*, **21**(2), 173–174.

Unwin, T. (1992) *The Place of Geography*. Longman Scientific and Technical, Harlow.

Unwin, T. (1995) Scholarship, applied industrial relevance and the historical geography of viticulture. In A. Jenkins and A. Ward (eds) *Developing Skill-based Curricula through the Disciplines: Case Studies of Good Practice in Geography*, SCED, Oxford, pp. 38–44.

School geography: the key questions for discussion

4

Tony Binns

Politics, economy and the school curriculum

Any attempt to evaluate school geography and suggest key questions for discussion will be strongly influenced by the particular experiences of the writer. This chapter is no exception. Rather than present a comprehensive overview of geography in schools, which in any case would be impossible, I have focused on specific issues which concern me as a parent, primary school governor, university lecturer and secondary teacher-trainer.

One point should be aired at the outset to provide the necessary context for what is to follow. Since the mid-1970s there has been an increasing, and possibly unprecedented, politicisation of education in England and Wales (see e.g. Binns, 1991, 1993, 1994; Bradford, 1994; Rawling, 1993). Many writers trace this process back to James Callaghan's Labour government in the 1970s, and specifically to the 'Great Debate' initiated by Callaghan's speech at Ruskin College, Oxford in October 1976 (*TES*, 1976).

The most frequently quoted elements of Callaghan's speech concern issues of 'efficiency' and 'accountability', which were later to become the 'buzz words' of the Thatcher Conservative administration during the 1980s. Callaghan observed that, 'We spend £6 billion a year on education. . . .

Geography into the Twenty-first Century, Edited by E.M. Rawling and R.A. Daugherty,
© 1996 John Wiley & Sons Ltd.

Parents, teachers, learned and professional bodies, representatives of higher education and both sides of industry, together with the Government, all have an important part to play in formulating and expressing the purpose of education and the standards that we need.' He continued to make the now familiar link between educational standards and economic performance, commenting that 'I am concerned on my journeys to find complaints from industry that new recruits from the schools sometimes do not have the basic tools to do the job that is required. . . . There seems to be a need for a more technological bias in science teaching that will lead towards practical applications in industry rather than towards academic studies' (*TES*, 1976). This might be seen as the cue for subsequent innovations, such as the Technical and Vocational Education Initiative (TVEI) of the 1980s and the more recent development of General National Vocational Qualifications (GNVQs), championed by John Major.

According to Rawling (1993), the so-called 'Great Debate', which followed Callaghan's speech, heralded the end of 'seven fat years' of innovative thinking and curriculum development in geography, and the start of a short transition phase leading to 'seven lean years' (1980–87) during the Thatcher regime, when all our efforts were diverted into lobbying and campaigning. The years 1975–95 have, indeed, witnessed great changes in attitudes to, and approaches within, education. With greater centralisation in educational decision-making, the role (and power) of local education authorities (LEAs) has been drastically reduced, teachers are no longer 'active partners' in education, but 'proletarianized deliverers of the National Curriculum', parents have become 'consumers', and schooling remains either the saviour and/or the scapegoat in relation to the fortunes of the national economy (Bradford, 1994). Geography has been unavoidably caught up in this political mêlée, though through constant vigilance and timely lobbying the subject in schools has shown remarkable resilience and innovation (Bailey, 1992).

Making a case for geography

I can well remember a discussion I had with Sir Keith Joseph, Secretary of State for Education, following his address to the Geographical Association (GA) at King's

College, London on 19 June 1985. I suggested to Sir Keith, albeit somewhat 'tongue in cheek', that geography really is the 'core' curriculum. If education is all about developing knowledge, understanding and skills in literacy, numeracy, global awareness and international understanding, I argued, then we already do all of this in geography. He looked at me pensively, hand on chin and then calmly said, 'please excuse me, I'm afraid I have to leave for a lunch engagement'! What more could I say?

In evaluating the position and relevance of geography in our schools, we must occasionally return to 'first principles' and ask ourselves 'what is the purpose of education and what is geography's peculiar contribution to the education process'? My comments to Keith Joseph do not actually seem too out of place when considered alongside the pronouncements of certain other writers. For example, Simon Jenkins, a former editor of *The Times*, wrote in 1988 that 'Geography is half of all education . . . . Geography stands as the one true science, the study of the human environment in all its manifestations. Its only equal in status is the study of human imagination and belief through literature. It makes the outside world make sense' (Jenkins, 1988). On a subsequent occasion, addressing the GA's Annual Conference, Jenkins mused, 'Surely geography was intellectually more central to education than, for instance, maths or science. Yours should be a core discipline' (Jenkins, 1992, p. 193).

The Geography National Curriculum Working Group in the late 1980s also initiated its deliberations from first principles, concluding in its *Interim Report* (DES, 1989, pp. 6–7) that geographical education should:

(a) stimulate pupils' interest in their surroundings and in the variety of physical and human conditions on the earth's surface;

(b) foster their sense of wonder at the beauty of the world around them;

(c) help them to develop an informed concern about the quality of the environment and the future of the human habitat; and

(d) thereby enhance their sense of responsibility for the care of the earth and its peoples.

I could personally work with these aims and, while there is always scope for criticism, I believe it would be difficult to improve on them. Unfortunately these aims were not subsequently seen through into a workable curriculum.

There have been numerous other attempts to 'stand back' and consider the rationale of geography in the education system and its particular aims and objectives. Bailey's opening chapter in *A Case for Geography*, for example, provides an admirable statement on this (Bailey, 1987, pp. 8–17). In the early years of this century, the GA's founding members also frequently considered geography's position in education and how the status and teaching of the subject might be improved. The influential Halford Mackinder, for example, urged teaching colleagues to treat geography 'as a discipline and not simply a mass of information' (Mackinder, 1903, p. 96), and he also encouraged consideration of progression in geography teaching. Others, such as Bryce, saw geography as 'the gateway to the physical sciences . . . the key to history and . . . the basis of commerce' (Bryce, 1902, p. 49). Like Jenkins some 90 years later, Bryce asserted that 'All branches of knowledge which have anything to tell us about the Earth more or less hinge into or are connected with geography' (Bryce, 1902, p. 50).

Thankfully, we have come a long way since T.G. Rooper, HMI, commented in 1901 that geography is, 'a dreary recitation of names and statistics, of no interest to the learner, and of little use except, perhaps in the sorting department of the Post Office' (Rooper, 1901). Since 1900, geographical education has steadily strengthened its position in British schools, colleges and universities. Boardman and McPartland, in an excellent series of four articles to celebrate the GA centenary, have traced the main developments in geographical education since the late nineteenth century (Boardman and McPartland, 1993a, b, c, d). From 'capes and bays' and rote learning of facts about world regions in the pre-Second World War period, through the impact of the 'quantitative revolution' in the 1960s and the innovative Schools Council curriculum projects of the 1970s and early 1980s, we have experienced most recently the rapid changes and challenges of the 1980s and 1990s, in the shape of the General Certificate of Secondary Education (GCSE), the National Curriculum and a vast catalogue of other educational innovations.

In spite of the relentless pace of educational innovation, particularly since the late 1970s, geography is very definitely 'live and kicking' in the schools of the 1990s. Numbers of young people taking geography remain buoyant. In 1994, for example, geography was the sixth most popular subject at both the GCSE and advanced (A) level (see Appendix 3). Whereas total A level entries increased by 13.4 per cent between 1989 and 1994, geography entries rose by 26.5 per cent in the same period. The popularity and perceived relevance of geography in schools have led to more students applying for higher education courses, which in turn produce geography graduates with apparently good levels of employment relative to many other subjects (Walford and Haggett, 1995, pp. 7–8).

Laying the foundations: geography in primary schools

It is not easy to suggest one single reason above others why geography is so popular in schools today. In primary schools in England and Wales, the onset of the Geography National Curriculum from September 1991 has done much to strengthen the position of geography in Key Stages 1 and 2 (5–11-year-olds). Earlier comments from HMI on primary geography suggested that there was considerable room for improvement. Between 1982 and 1986, HMI undertook a monitoring exercise involving the inspection of 285 primary schools. Their published report comments, 'Overall standards of work in geography were very disappointing: in only one-quarter of both infant and junior schools and departments were they satisfactory or better' (HMI, 1989, pp. 11–12). HMI observed that where the topic work approach was used, 'there was tendency for geography to lose its distinctive contribution and to become a vehicle for practising skills related to language and art' (HMI, 1989, p. 12). Work on distant places was generally weak: 'the almost total absence of a national and world dimension to the work in many cases highlighted the need for schools to consider a broader perspective' (HMI, 1989, p. 12). The same HMI survey also drew attention to the absence of specialist geography teachers in many primary schools.

Since September 1991 there has been some early evidence to suggest that the quality of primary geography is gradually improving, but in some schools there is still a long way to go. Prior to the onset of the National Curriculum, commentators often suggested that geography in primary schools, while present, was frequently 'disguised' in topic work (Catling, 1992). If nothing else, the new statutory requirements should make curriculum planning easier and bring geography 'out of the shadows', making it much more explicit in primary schools. In-service training courses and the emergence of a wide range of resources, not least the GA's primary magazine *Primary Geographer*, have generated much interest among teachers and seem to be having a positive impact on school work. Primary membership of the GA has grown spectacularly since the introduction of the National Curriculum, such that half of its 11 000 strong membership is now (1995) from the primary sector.

The past few years have not been easy for primary school teachers, and the initial Geography National Curriculum Order, with no fewer than 5 attainment targets and 183 statements of attainment, proved to be one of the most difficult for primary schools to translate into practice (Lloyd, 1994, pp. 7–9). This is understandable when one realises that in addition to a shortage of primary geography specialists, most primary teachers unlike their secondary colleagues, have to become familiar with the statutory requirements of all the National Curriculum subjects. Lloyd (1994) found much more evidence of collaborative planning in primary schools, particularly when there were effective geography co-ordinators.

The main weakness in primary geography still seems to be in teaching about places beyond the local area. However, resources for distant place work are improving all the time, for example in the shape of photo-packs and television series. Visits and in-service courses at locations such as the Commonwealth Institute and the Museum of Mankind are also proving popular. Another interesting development in primary schools has been the forging of school links at home and overseas. The junior school where I am a governor has recently established a link with a school in The Gambia, and the stimulating contacts and material generated have now been developed into a term-long module on Africa for Year 6 children.

The further strengthening of geography in primary schools is vital for the future popularity of the subject in the secondary and tertiary sectors. This is now even more significant with the revised post-Dearing National Curriculum where, with the decision to confirm geography as an optional subject in Key Stage 4 (14–16-year-olds), six of the nine years of statutory geography teaching are now undertaken in the primary sector.

Building on the foundations: geography in secondary schools

With the introduction of the Geography National Curriculum from September 1991, secondary teachers should, theoretically, now have a much better understanding of what children have been taught in their 'feeder' primary schools. In reality, however, although there is more 'common ground' in primary geography, individual schools will make use of the flexibility in the curriculum to choose their own case study material to support their teaching. The importance of good liaison at the primary–secondary interface remains, but in most cases liaison is less effective than it could be. Although many secondary schools have links with 'feeder' primary schools, these are generally 'whole school' links and not subject specific. I firmly believe that the further strengthening of links between primary and secondary geography teachers could be extremely valuable, perhaps fulfilling an important in-service training role for non-specialist teachers, at a time when LEA in-service training provision is steadily diminishing.

Innovation in secondary school geography

Geography's present strength in secondary schools is undoubtedly due, at least in part, to three curriculum development projects sponsored by the Schools Council before it was abolished by the Thatcher government in 1984. Two of these projects were started in 1970 and focused on 14–16-year-olds, Geography for the Young School Leaver (known as GYSL or the Avery Hill Project), and Geography 14–18 (the Bristol Project). A third project, the Geography 16–19 Project, aimed at the 16–19 age group, started in 1976. These three curriculum projects injected new life into school geography and encouraged young people to think, question and analyse issues and problems. This led to more varied teaching strategies such as group work, role plays and simulations.

The projects also played an influential part in initiating the development of new resources, such that the resources for teaching secondary-level geography in the 1990s are among the best available from publishers. Compared with the 1960s, recent geography textbooks probably have much less actual text, but they are colourful and exciting, involving students in decision-making and in evaluating arguments and data. In addition there is a wealth of other supportive material for geography teaching in the shape of software packages, video, television and radio series, photo-packs and much more.

While school-based curriculum development and innovation have probably been constrained recently by an over-prescriptive National Curriculum (Rawling, 1995), the approaches and resources associated with these earlier curriculum development projects have 'rubbed off' on many other more traditional syllabuses. The Geography 16–19 Project A level is now the most popular A level syllabus, experiencing a meteoric rise since the mid-1980s. With increasing numbers of students now entering higher education with the 16–19 experience of A level, university and college teachers are having to adapt their own first-year courses and teaching strategies, to bring them more into line with their students' earlier experience. This innovation in higher education teaching is to be welcomed, and is likely to accelerate as 'quality audit' inspections become as regular a feature of higher education as the now well-established research assessment exercises. Just as there is a need for strengthening links across the primary–secondary interface, so there is an equally important need for developing similar links across the secondary–tertiary interface. Remarkably, few higher education teachers have a detailed understanding of geography teaching at A level, let alone of the pre-16 age group. Some would cynically suggest that interest in school geography among higher education teachers only increases when their student applications are falling!

Geography, humanities and cross-curricular links

Although geography is currently in a strong position in secondary schools, during the late 1970s and early 1980s it seemed at times as if geography might lose its identity in schools and be subsumed within 'humanities'—whatever that might mean. Bailey (1989) has shown how key documents from HMI in 1977 and 1980 and from the DES in 1980, had very little, if anything, to say specifically about geography (HMI, 1977, 1980; DES, 1980). In the period leading up to Sir Keith Joseph's GA address in 1985, there was a strong feeling within the Association that the Secretary of State firmly believed that geography and history were inextricably entwined and should be delivered together in schools under the humanities 'umbrella'. There was a deep sense of relief when this issue did not figure prominently in Sir Keith's address, and his seven questions actually gave the GA an opportunity to reassert its case for geography to be taught as a subject in its own right (Bailey and Binns, 1987). In fact the GA went further, to stress that geography's links with other subjects were as strong, if not stronger, than they were with history. In particular, the GA emphasised geography's long-standing links with science and a Geography–Science Working Group was established by the Association to audit existing geography–science links in schools and to encourage further collaboration. There is evidence that such collaboration is increasing, not least because as a core subject in the curriculum, science has more teaching time and resources allocated to it, and geographers might therefore gain from strengthening links with their science colleagues (Adamczyk et al., 1994).

Geography still continues to be delivered as an element of 'humanities' in many schools, particularly in the first two years of secondary education. While some schools have undoubtedly given careful thought to the underpinning philosophy and coherence of integrated humanities pro-grammes, in many other institutions there is no clear rationale for humanities, other than as a logistical and timetabling convenience. HMI observations support these comments and are generally critical of many humanities courses. For example, in a survey of humanities teaching in 26 schools, HMI found 'there was a lack of consensus about the nature of "humanities" (and). . . The uncertain curriculum structures of many humanities courses, and the pupils' fragmented and often partial experience of the contributing humanities subjects meant that the general standard of work when taken as a whole was only fair, and needed to be improved' (HMI, 1991, p. 1). With the introduction of the subject-specific National Curriculum, it was thought likely that humanities courses would become less important and perhaps eventually disappear altogether. This does not seem to be happening, but schools will nevertheless have to demonstrate during inspections that statutory National Curriculum requirements for each subject are adequately covered.

Some secondary schools have recognised that geography can play an important role in many other aspects of the curriculum. In addition to science and history, there is scope for more effective collaboration between geography teachers and their colleagues in, for example, art, music, English, drama, modern languages, mathematics, design and technology. In the period immediately following the 1988 Education Reform Act, some interest was shown by the Curriculum Councils in England and Wales in cross-curricular themes, including environmental education, economic and industrial understanding and citizenship. These cross-curricular themes were supposed to 'foster discussion of questions of values and beliefs; they add to knowledge and understanding and they rely on practical activities, decision making, and the inter-relationship of the individual in the community' (NCC, 1989, p. 6). Geography's potential contribution to environmental education, for example, is considerable and was recognised by teachers and others long before the coming of the National Curriculum. The original geography National Curriculum Order actually included an attainment target, rather awkwardly labelled 'environmental geography'. While the idea of cross-curricular themes was stressed by the Geography National Curriculum Working Group and generally welcomed by teachers, it was not promoted by government ministers, and the extent to which schools have developed and integrated these themes into their curricula is extremely variable. Furthermore, the revised National Curriculum effective from 1 August 1995, has very little to say about cross-curricular themes. It seems likely that even fewer schools will take them seriously in the future, though the 20 per cent of school time 'freed up' by the revised curriculum in theory actually provides greater opportunities for cross-curricular themes to receive more attention.

Darkening clouds: Key Stage 4 and GNVQ

Undoubtedly a significant landmark in post-war geographical education was the announcement in March 1987 from

Kenneth Baker, Secretary of State for Education, that geography was to be a 'foundation subject' in the new National Curriculum, and as such would be taught to all children in England and Wales during the period of compulsory schooling from 5 to 16. This was recognised as a great achievement for the subject. However, Key Stage 4 (14–16) geography, in the shape of revised GCSE syllabuses, lost its short-lived compulsory status when it was realised that the curriculum was heavily overloaded and schools would have difficulty delivering core and foundation subjects to all children between 5 and 16. The loss of compulsory status for geography at Key Stage 4 is regrettable, and means, in effect, that we have returned to the subject being one of a range of options at GCSE, as it was in most schools before the National Curriculum. However, what is now considerably different, is that the core subjects—English, maths and science (together with Welsh in certain Welsh schools)—have been significantly strengthened by the statutory requirements, and unlike other subjects will be assessed by national tests at the ages of 7, 11 and 14. There is a danger that the non-core subjects, including geography, will be squeezed as greater attention is given to the core triumvirate.

A further possible threat to the popularity of geography in secondary schools, with likely 'knock-on' effects for higher education, is the rapid development of the so-called 'vocational pathway', championed by Prime Minister John Major (himself a 'young school-leaver'!) and backed by considerable funding and publicity. General National Vocational Qualifications (GNVQs) were first announced in the White Paper, *Education and Training in the 21st Century*, published by the Conservative government in May 1991. John Patten, Secretary of State for Education, commented on 5 April 1993 that 'GNVQ's are not a soft option. They will stand alongside academic qualifications on their own merits. . . . In the longer term I would like to see GNVQs developing into a mainstream qualification for young people, catering for as much as half the age group' (NCVQ, 1993, p. 3). With schools being encouraged to develop vocational courses for 14–16-year-olds (Key Stage 4) as well as for post-16 students, vocational options could provide stiff competition for other optional subjects, such as geography, in the post-14 age group. Parents and students may perceive the new vocational courses as being more relevant in terms of future career prospects, although in fact various studies have shown that geography fares well relative to other subjects in terms of the employability of its students (Walford, 1991).

On a more positive note, there would be significant opportunities for geography to contribute to GNVQ courses in, for example, land-based and environmental industries, leisure and tourism, construction and the built environment, information technology (IT) and manufacturing. There will be some, however, who will argue that geography could lose its intellectual purpose if it becomes too closely associated with vocational education (Butt, 1994, p. 183). Whichever point of view is taken, the development of the vocational 'pathway' is gaining momentum and likely to be given further impetus by the merging of the Departments for Education and Employment in mid-1995. The situation will need to be monitored carefully, both in respect of the potential involvement of geography and in the effects on geography recruitment at GCSE, A level and degree level.

A level: maintaining the 'gold standard'

There has been considerable debate over many years concerning the nature of educational provision and curriculum content in the post-16 non-compulsory sector. Conservative governments since 1979 have regularly stated a commitment to maintaining A levels as the 'gold standard' of academic achievement and the main qualification for entry into higher education. However, there have been many calls from education practitioners for a broadening of the post-16 curriculum, possibly on the lines of the Scottish 'Higher' system and the International Baccalaureate, in which five or more subjects are studied compared with the usual three at A level. The Higginson Report (DES, 1988a) presented the conclusions of a government-initiated working group under the chairmanship of Gordon Higginson, Vice Chancellor of Southampton University. Higginson recommended a move towards sixth formers taking five 'leaner and tougher' A level syllabuses which would 'stimulate intellectual curiosity, encourage general investigation and communication skills and develop those qualities which are picked out in the notion of a trained mind' (DES, 1988a, p. 12). Having established this committee, the Secretary of State and the government (most notably Prime Minister, Margaret Thatcher) refused to implement Higginson's recommendations commenting that 'The Government endorses the general aim of broadening A level students' programmes of study, but does not accept the Committee's proposals for achieving that objective through a five subject programme incorporating leaner A level syllabuses' (DES, 1988b). Instead the government advocated AS (Advanced Supplementary) courses, 'as the key to achieving greater breadth' (DES, 1988b). AS levels were first proposed in 1984 to broaden the sixth-form curriculum, with a series of 'rigorous and intellectually demanding' courses taking up about half the time of average A level courses. Students were encouraged to take a mix of A and AS levels, the latter crossing the arts–science boundary wherever possible. However, the adoption of AS levels in schools and colleges has been patchy, largely because of inadequate resourcing. Their continuing existence and future development alongside the new and better resourced GNVQs are also doubtful.

Recent discussion relating to A and AS levels has concerned the identification of a core of study, which in future must be incorporated into all syllabuses, and should comprise about one-third of an A level syllabus and two-thirds of an AS syllabus (Butt and Lambert, 1993). The 'core', produced by the School Curriculum and Assessment Authority (SCAA), is purposely rather general and will have needed 'fleshing out' in the new syllabuses starting in September 1995, for first examination in 1997. According to SCAA, 'the core has been developed to provide a sound basis for the study of geography whilst allowing flexibility of approach and emphasis' (SCAA, 1994, p. 1). The idea of a 'core–option' model is nothing new in geography, as the successful Geography 16–19 syllabus has used this approach from its inception in the early 1980s.

The structure and character of the whole 16–19 educational system have been thrown into further uncertainty by the establishment of the Review of 16–19 Qualifications to be undertaken by Sir Ron Dearing during 1995/96 (DfE, 1995). Sir Ron's brief, and subsequent elaboration of this, referred to the need to broaden the 16–19 curriculum, to consider ways of promoting parity of esteem between vocational and academic courses and to maintain the rigour of A levels. It remains to be seen whether the resulting Dearing proposals due at Easter 1996, will strengthen or weaken the position of geography at this level.

Geography in schools: towards the next millennium

In trying to draw together various strands in school geography and looking to the future, I would like to pose two key questions, more to provoke further thought rather than to offer a definitive answer:

1. In the light of significant developments in education policy, such as the loss of geography's compulsory status at Key Stage 4 and the introduction of GNVQs, have we reached the high-water mark in the popularity of school geography, as reflected in examination entries at GCSE and A level?
2. If this is the case (and even if it is not), how can we further strengthen the position of geography in our schools and improve the quality and relevance of geographical education?

The years since 1975 at least have demonstrated how geography in the education system of England and Wales is a prime case of 'bottom-up' development, where the success of the subject in higher education has largely stemmed from dynamic, challenging and popular geography teaching in our schools. In sharp contrast, school geography in the USA, for example, has a much weaker identity and is frequently taught as part of social studies. Concern about the quality of geographical education in US schools led to the establishment in 1988 of the Geographic Education National Implementation Program (GENIP), sponsored by a number of geographical bodies such as the National Geographic Society and the Association of American Geographers. The National Geographic Society has also initiated Summer Institutes in Geography for teachers, encouraged State Geographic Alliances and been instrumental in promoting an annual National Geography Awareness Week. However, before these various initiatives, geography's low profile in US schools was undoubtedly a factor in the closure of geography departments at some prestigious universities, most notably Harvard, Yale and Northwestern (Walford and Haggett, 1995).

The comparison between UK and US experiences reveals clearly the important relationship between the strength of geography in schools and higher education institutions.

Andrew Goudie, as President of the GA, echoing the sentiments of Professor S.W. Wooldridge almost 40 years earlier (Wooldridge, 1955), lamented the lack of involvement of higher education geographers in the GA, which he suggested was due to factors such as academics being 'increasingly driven by the desire to do research and have less time for teaching and education', while at the same time the GA 'has become more concerned with Education per se than with Geography'. He concluded that 'A chasm has developed between those who teach at school and those who teach in universities' (Goudie, 1993, pp. 338–339).

If we are to ensure the future well-being of the subject, links between the various sectors of the education system must be further developed. There must be greater mutual respect and understanding between geographers in different sectors of the education system. The Madingley Hall seminars in the early 1960s were a prime example of effective 'trickle down' development, when influential academics, such as Peter Haggett, inspired young teachers and set an agenda (much needed at the time) for the future of geographical research and teaching. Since then, as we have seen, innovation in geographical education has been more 'bottom-up', generated by the various curriculum projects, new teaching approaches and some impressive resources. Unlike the quantitative revolution, with its models, scientific method and statistical techniques, recent developments in academic geography, such as the 'new cultural geography' and 'post-modernism', have thus far made little impression on geography in schools. The Council of British Geography (COBRIG) seminar in July 1994, of which this book is a tangible product, served to revive the all-important dialogue between a diverse range of geography practitioners. The seminar also clearly demonstrated the importance of COBRIG itself, founded in 1988 to provide a unique forum for debate and action among its 9 constituent organisations from England, Scotland and Wales.

In addition to the need for further cross-sector collaboration in UK geography, we must also work hard to develop our international links. Whilst British geographical education has a high international reputation, we are at times somewhat reluctant to take the lead in international matters. The International Geographical Union (IGU) and the GA's International Committee are just two of a number of bodies which are striving to bring together geographers from different countries, to share experiences and to speak with one voice when necessary. The IGU's *International Charter on Geographical Education* (IGU, 1992) provides a valuable statement on the contribution of geography to education, stating emphatically that 'geography is both a powerful medium for promoting the education of individuals and a major contributor to International, Environmental and Development Education' (p. 7). This charter provides useful ammunition for justifying the place of geography in the school curriculum.

But if we are to maintain the vibrancy, topicality and relevance of geographical education in schools well into the twenty-first century, I believe there are two further, closely related, areas which need constant attention. These are the importance of direct experience and the recognition that for many of us geography is about places. Travel and fieldwork together generate an empathy with places and the peoples and environments which have shaped these places. I am writing this chapter at a time when geographical fieldwork in our schools (and indeed also in higher education) is once again under attack from a combination of factors including concerns about safety, cost, curriculum time and school logistics. Direct experience, whether in the school grounds, the local area or more distant places, is as vital to the future success of school geography as it is to teaching and research in higher education. Furthermore, I feel sure many would agree when I say that the value of residential fieldwork in academic and social terms far outweighs the required inputs of time and finance. We must be vigilant, and articulate a strong case for the presence and enhancement of fieldwork and direct experience as an integral element of all good geographical education.

Finally I would like to consider what is a central issue in geography and geographical education, the 'question of place', which coincidentally is also the focus of Ron Johnston's chapter (5). I share Johnston's views about the significance of place in geography and, in the context of the present chapter, places in school geography. During the formulation of the Geography National Curriculum there was a lively debate about the extent to which 'places' should figure in the statutory requirements. Whether we like it or not, the popular image of geography is that it is about places, and a number of widely publicised reports during the 1980s revealed a widespread weakness in place knowledge among the general population (see e.g. Gallup Organisation, 1988). While I am firmly against the statutory prescription of specific countries for study in schools, since it could lead to a 'partial' view of the earth and its peoples, I equally firmly believe that young people are fascinated with places, and it is this fascination, sensitively nurtured and stimulated by the teacher, which is the essence of good geographical education.

Conclusion

A lot of water has flowed under the bridge in the 10 years since some of us gathered in June 1985, somewhat apprehensively, to hear Keith Joseph's thoughts on geography in schools. Since then, the educational world has become ever more political, with one innovation following another, invariably introduced with inadequate time and resources and frequently without the goodwill of those whose duty it has been to ensure grassroots implementation. But those of us concerned with geographical education should be reassured that we have done much more than just keep our heads above the water. We have, I believe, become adept at playing the politician's game, and the geographical fraternity has responded on many occasions, usually with one voice, to a wide range of pronouncements from Westminster concerning geography and a variety of educational issues which might impinge on the subject, either directly and indirectly. The success of our work is amply demonstrated by the popularity of our subject at GCSE, at A level and in higher education.

But as I have already indicated, there are a number of significant recent developments which must temper any desire for complacency. We must continue to be both vigilant and assertive if we are to maintain geography's currently strong position in our schools as we approach the twenty-first century. There will be many future occasions when we will have to reassert our case for geography. Above all, we will constantly need to convince politicians, parents, headteachers, teachers, governors and of course our students, that geography is exciting, dynamic and above all *relevant* to the needs of our society, economy and the individual. I believe we have a strong case to offer and we are starting from a sound base. COBRIG is well placed to bring together our different skills and perspectives and to co-ordinate action.

Acknowledgements

I would like to thank Eleanor Rawling and Rex Walford for their helpful comments in writing this chapter.

References

Adamczyk, P., Binns, T., Brown, A., Cross, S. and Magson, Y. (1994) The geography–science interface: a focus for collaboration. *Teaching Geography*, **19**(1), 11–14.

Bailey, P. (1987) What are the geographer's contributions? Geography in the curriculum from 5 to 19. In P. Bailey and T. Binns, *A Case for Geography*, the Geographical Association, Sheffield, pp. 8–17.

Bailey, P. (1989) A place in the sun: the role of the Geographical Association in establishing geography in the National Curriculum of England and Wales, 1975–89. *Journal of Geography in Higher Education*, **13**(2), 149–157.

Bailey, P. (1992) Geography and the National Curriculum. A case hardly won: geography in the National Curriculum of English and Welsh schools, 1991. *Geographical Journal*, **158**(1), 65–74.

Bailey, P. and Binns, T. (1987) *A Case for Geography*. The Geographical Association, Sheffield.

Binns, T. (1991, 1993, 1994) Geography and education: A UK perspective. *Progress in Human Geography*, **15**(1), 57–63; **17**(1), 101–110; **18**(4), 541–550.

Boardman, D. and McPartland, M. (1993a) A hundred years of geography teaching. Building on the foundations: 1893–1943. *Teaching Geography*, **18**(1), 3–6.

Boardman, D. and McPartland, M. (1993b) A hundred years of geography teaching. From regions to models: 1944–1969. *Teaching Geography*, **18**(2), 65–68.

Boardman, D. and McPartland, M. (1993c) A hundred years of geography teaching. Innovation and change: 1970–1982. *Teaching Geography*, **18**(3), 117–120.

Boardman, D. and McPartland, M. (1993d) A hundred years of geography teaching. Towards centralisation: 1983–1993. *Teaching Geography*, **18**(4), 159–162.

Bradford, M. (1994) *Diversification and Division in the English Education System: Towards a Post-Fordist Model*. SPA Working Paper 29, Dept of Geography, University of Manchester, Manchester.

Bryce, J. (1902) The importance of geography in education. *The Geographical Teacher*, **1**(2), 49–61.

Butt, G. (1994) Geography, vocational education and assessment. *Teaching Geography*, **19**(4), 182–183.

Butt, G. and Lambert, D. (1993) Modules, cores and the new A/AS levels. *Teaching Geography*, **18**, 180–181.

Catling, S. (1992) Issues for the future of primary geography. In M. Naish (ed.), *Geography and Education*, Institute of Education, London, Ch 1, pp. 9–33.

DES (Department of Education and Science) (1980) *A Framework for School Education*. HMSO, London.

DES (1988a) *Advancing A levels* (Higginson Report). HMSO, London.

DES (1988b) Parliamentary answer by the Secretary of State outlining the government response to the report 'Advancing A levels', 7 June, London.

DES (Department of Education and Science and the Welsh Office) (1989) *National Curriculum Geography Working Group, Interim Report*. DES, London.

DfE (Department for Education) (1995) Terms of reference for Sir Ron Dearing's Review of the 16–19 Qualifications framework (published with press release April 1995).

Gallup Organisation (1988) *Geography: An International Gallup Survey* (Report for the National Geographic Society, USA). Gallup, Princeton NJ.

Goudie, A. (1993) Schools and universities—the great divide. *Geography*, **78**(4), 338–339.

HMI (Her Majesty's Inspectorate) (1977) *Curriculum 11–16. Working Papers by HM Inspectorate: a Contribution to Current Debate*. HMSO, London.

HMI (1980) *A View of the Curriculum. Matters for Discussion 11*. HMSO, London.

HMI (1989) *Aspects of Primary Education: The Teaching and Learning of History and Geography*. HMSO, London.

HMI (1991) *Inspection of Humanities Courses in Years 5–9 in 26 Schools (Sept. 1989–May 1990)*. DES, London.

IGU (International Geographical Union) (1992) *International Charter for Geographical Education*. Commission on Geographical Education, IGU.

Jenkins, S. (1988) Geography puts on its glad rags at last. *Sunday Times*, 3 April.

Jenkins, S. (1992) Four cheers for geography. *Geography*, **77**(3), 193–197.

Lloyd, K. (1994) Place and practice: the current state of geography in the primary curriculum. *Primary Geographer*, **17**, 7–9.

Mackinder, H.J. (1903) Report of the discussion on geographical education at the British Association meeting. *The Geographical Teacher*, **2**(3), 95–101.

NCC (National Curriculum Council) (1989) *Curriculum Guidance 3: The Whole Curriculum*. NCC, York.

NCVQ (National Council for Vocational Qualifications) (1993) *GNVQ Information Note*. NCVQ, London.

Rawling, E. (1993) School geography: towards 2000. *Geography*,

78(2), 110–116.

Rawling, E. (1996) School-based curriculum development in secondary geography; the impact of the National Geography Curriculum. In A. Kent, D. Lambert, M. Naish and F. Slater (eds) *Geography in Education.* Cambridge University Press, Cambridge.

Rooper, T.G. (1901) On methods of teaching geography. *The Geographical Teacher,* 1, 4–10.

SCAA (1994) *A/AS Subject Core.* SCAA, London.

TES (Times Educational Supplement) (1976) What the PM said. *TES,* 22 October, p. 1.

Walford, R. (1991) Careers for geographers: what prospects for the 1990's? *Geographical Journal,* **157**(2), 199–206.

Walford, R. and Haggett, P. (1995) Geography and geographical education: some speculations for the twenty-first century. *Geography,* **80**(10), 3–13.

Wooldridge, S.W. (1955) The status of geography and the role of field work. *Geography,* **40**(2), 73–83.

Part B

Geography in higher education

5   A place in geography

R.J. Johnston

An interest in places

People are attracted to the study of geography for a range of reasons; so too are those drawn to an occupation as a geographer. For many, the initial fascination is aroused through an interest in places, their characteristics and variety. That appeal may be stimulated by direct experience of one or more places, by the study of documentary evidence such as maps, or, increasingly, by exposure to places through visual media. Whatever the origin of the curiosity, however, the goal becomes the same: to appreciate the diversity which characterises the earth's surface, and to understand its origins.

As members of an academic discipline, geographers have tackled this goal in a number of ways, as illustrated by historical studies of the discipline over both long (Livingstone, 1992) and short (Johnston, 1991a) time-scales. Because its educational role has long dominated its existence, the discipline has been particularly influenced by perceived needs for geographical materials—as illustrated by Sitwell's (1993) analysis of the nature and publication chronology of the 'special geographies' which characterised geographical 'texts' for several centuries, books which informed both a general and an educational audience of the diverse nature of the earth's myriad places.

Geography into the Twenty-first Century, Edited by E.M. Rawling and R.A. Daugherty,
© 1996 John Wiley & Sons Ltd.

In common with all academic disciplines, the search for understanding of complex reality has induced some excursions up culs-de-sac; what for a time appeared to offer exciting new opportunities for study have then been discarded, by most if not all adherents, although some aspects of what has been done may be incorporated within further attempts at restructuring the discipline. In this chapter, I briefly outline the nature of one cul-de-sac, before turning to an appreciation of the role of place in society and hence its importance to the practice of geography, especially human geography.

A scientific cul-de-sac?

Geographical practice between the mid-1950s and the mid-1970s was dominated by what is often referred to as the 'theoretical and quantitative revolution' (Johnston, 1991a). This occurred in response to a number of factors, including similar changes in other disciplines, a general favouring of science within society as a means of explaining and thus managing complex realities (both 'natural' and 'human'), the increased importance placed on research within universities, and the general growth of geography as an academic discipline (based on the number of students opting to read for geography degrees), which allowed research specialisms to develop. Geographers, whatever the subject matter on which they focused their attention, were increasingly attracted to 'scientific' study, and the new generations of postgraduate researchers who joined them were rapidly seduced by this apparently more prestigious and valuable approach to their discipline (on which see Gould, 1979, 1985). The particularities of places—often termed regions—could be explained as the operation of general processes in specific contexts (see Berry's, 1964, programmatic statement).

Fundamental to this approach was a belief that all which is observed can be explained as a consequence of the operation of general laws. The geographer's task, therefore, was to uncover those laws through application of the methodology of science, which involves creative hypothesising and rigorous testing, using robust and replicable (quantitative) techniques. It was accepted that this would not be straightforward, because the operation of something as complex as the environment or a modern society involved many laws operating simultaneously, interacting on each other in diverse ways. But that added to the attraction of the challenge.

Within this overall framework, many different approaches have been tried: most have at least partially failed to provide convincing explanations (on which predictions and planning can be based). In physical geography, successful scientific understanding of the operations of nature is extremely difficult to achieve, given the complexity and scale of the task. In human geography, on the other hand, many came to the conclusion that the framework itself is fundamentally flawed—although much that was tried within that framework had lasting value, as with many of the technological developments which facilitated much improved descriptions of the world we live in.

Place and scientific 'failure'

In both human and physical geography, failure to explain and predict was in part a consequence of a lack of appreciation of the importance of place, though in very different ways. 'Place' is a very commonly used word, but rarely carefully specified, and its usage has created many problems for human geographers (Johnston, 1991b). Sitwell (1993, p. 4) provides a suitable definition, based on earlier work by Robinson (1973). A place is

> an area on the surface of the earth whose approximate location and extent is known, as shown by the fact that *people have a name for it*. The emphasis is added to underline that fact that the criteria for establishing the existence of a place are subjective; given the arbitrary nature of language, there is no guarantee that the criteria will be understood and so used in the same way by different people.

Places can thus vary greatly in their size—the Atlantic Ocean is a place on such criteria, as is Swindon. Sitwell (1993, pp. 4–5) establishes a minimum scale threshold for a place within geographical work, however:

> Places are larger than a human being by at least an order of magnitude. Examples are provided by a cluster of houses, a lake, a wood, a hill. What these have

> in common is the fact that they are so large that we
> cannot take them in at a glance unless we are removed
> from them to a considerable distance—which is to say,
> not until they have been, in terms of our literal perception
> of them, reduced in scale.

This definition is as valid for physical as for human
geographers.

Physical geography

For physical geographers, a place is a defined area, with
both substantial internal homogeneity and a separate identity
from its neighbours, on predetermined criteria. Their goal is
to understand its physical characteristics, through the
application of general scientific laws which control environmental processes operating there. This is usually a complex
task, for two reasons. First, most places have complex
environmental characteristics, with a large number of
processes interacting together in ways that may not be
exactly reproduced elsewhere. Second, no place is a closed
system, isolated and insulated from what is happening in
others: interactions across place boundaries thus add further,
very extensive, complexities to the subject matter which is
the focus of attention.

This argument can be illustrated from the study of
climatology. A place—climatic region—can be defined on
specific criteria, and work is undertaken to explain its
climate as consequences of the interaction of general
atmospheric processes with topography and other local
circumstances. Beyond a very broad level of generalisation,
however, those interactions are extremely difficult to model
and appreciate, hence the great difficulties of weather
forecasting in many parts of the world. The complexity of
the interactions, and the difficulty of appreciating them, has
been emphasised in recent years with developments in the
mathematics of complex systems. Chaos theory, for example,
shows how slight changes in the initial conditions for a
process can result in very different outcomes (often very
soon), thereby confirming Curry's (1962) prescient representation of climatic change as a random series (see Johnston,
1989a). Furthermore, similar complex interactions with
unexpected (if not, in the ultimate sense, unpredictable)

consequences are happening concurrently in other places too, with effects that spill over the place borders, thereby further complicating the operation of general processes in others. The impact of the eruption of Mt St Helens on weather around the world illustrates this; and, to change the scale, a tree which is blown over in a storm may fall into a streambed and alter the direction of water flow.

Physical geographers face very exacting problems as they seek explanations of how environmental processes create particular suites of landforms, therefore. Those problems do not invalidate use of the scientific procedures; they just make them very difficult to apply in context (the so-called 'real world'). Nature, in all its diversity and apparent unpredictability, results from the operation of unchanging fundamental scientific laws. Gravity is invariant (or virtually so) within the earth's atmosphere, for example, and even though it cannot be apprehended directly, nevertheless, a law has been deduced which predicts its impact in certain conditions. The same is true of many other laws. It is when the processes that they generate interact in particular situations that the difficulties arise. There are so many ways in so many different circumstances, that the permutations become massive. Understanding will slowly improve, but achieving it will probably call for much greater scientific expertise than characterises geographers at present.

Human geography

Human geography differs from physical geography in a variety of ways, but perhaps the most fundamental is that its subject matter is not the product of invariant scientific laws. There are two types of scientific law (Johnston, 1989b). The classification of phenomena into categories which have common characteristics (resulting from the operation of particular processes) involves the use of *membership laws*. Once those have been established, then *functional laws*—of cause and effect—can be developed to account for behaviour (including spatial outcome patterns) within each class of phenomena. Both are fundamental to physical geography: neither is to human geography.

Central to the argument that membership laws are not applicable in human geography is the difference between uniqueness and singularity. A unique event may be repeated;

a singular event can never be replicated (except 'artificially'). Unique events—such as the October 1987 hurricane in southern England—result from the operation of general functional laws in a specific context, even though that context (the contingent circumstances of the event) may never recur. Singular events, on the other hand, result from decisions which are not generalisable.

Singularity is characteristic of human geography, therefore, because individuals cannot be represented as just a collection of general mental processes brought together in a particular person. We are each more than a collection of living cells, because we have independent powers of reason that we apply in particular contexts. Those contexts are ever-changing, because we learn continually—and so does everybody else: the world in which we operate is never the same from one moment to the next. Of course, we share many characteristics in common with other people—through inheritance and, more importantly socialisation in the same culture—and because of this some of our actions are 'predictable'. (As, for example, with the expectation that people of a certain socio-economic status in Great Britain will vote for the Conservative Party; but many of them do not!) But we are not machines which react to stimuli in pre-programmed ways. We govern our own actions, albeit in circumstances which are usually not of our making and which therefore constrain as well as enable application of our decision-making powers. Those circumstances are shared with others, but only in part, because we have to interpret them. Our interpretations are much influenced by our socialisation into one or more cultures, and by our contacts with others at the time, but they are individual to us. How we act, therefore, is singular, even though any particular act may have much in common with what others do. It is not explicable as a realisation of a general law in operation. Physical geographers may be able to predict what would happen if all other laws than those they were studying were held constant—what they call the *ceteris paribus* situation: human geographers cannot, because all other things are never the same—what Hagerstrand called the *ceteris absentibus* situation.

This may appear to create a situation of some despair, suggesting that human geographers can never do more than describe events which are singular in space and time. This is not so, however, because of the nature of the learning process identified above. We are created with mental faculties and could, presumably, learn to survive using them and with little or no interaction with any other individuals. But this never happens; we learn from and with others (from, early in life; with, thereafter)—most often in *places*. Thus places are crucial to the development of individuals, and hence to their understanding. People become what they are because of where they are, and places in turn become what they are because of the people who live in them.

In this context, a place is much more than a collection of animate and inanimate objects. It is the locale of a particular culture, providing knowledge about the world which is drawn upon as residents face stimuli, and which is added to as they observe the consequences of their responses to those stimuli. Their knowledge is shared with others in the place, who have institutional arrangements (notably educational systems) for its transmission.

Place is crucial to human geographers, therefore, because it is the individuals' learning context, the arena in which they learn to be humans and then act as such. During our lives, most of us will live in many places, often simultaneously or virtually so. Many occupy two places daily, for example, a place of work and a home; we learn in each, and transfer (to a greater or lesser extent!) the fruits of our learning in one place to our activities in the other. Thus we change as we learn through coming into contact with others in other places, and they change too: we are continually involved in cultural evolution.

But are there no laws for human geographers?

The difference between human and physical geography described above could be interpreted as indicating that they have nothing in common—other than that their subject matter is spatial variation across the earth's surface. But this is far from the case; from one vantage point they have much in common, not least because one of the main subjects of human interpretation is the environment. (Indeed, much physical geography is, in effect, involved with testing the veracity of such interpretations.)

Physical geography is a discipline which seeks explanations of phenomena as exemplars of general laws operating in particular circumstances. The laws do not vary, but the circumstances—the contingent conditions—do. In other words, the characteristics of a place influence how processes operate there, and so contribute to either the maintenance of, or change in, those characteristics. Gravity is ever present, and causes water to flow downhill, but how that flowing water affects the development of the environment will depend on its (micro and macro) characteristics. Thus there are general mechanisms producing unique events. Human action is one of the contingent conditions which stimulates those mechanisms, as with the use of fire to change a landscape. Once we know that a fire has been started in a particular place, whose other characteristics (topography, weather at the time, etc.) are also known, then we can predict with some accuracy what the fire will do; but we can never predict when the fire will be started, and in what particular local conditions. (For more on this realist approach to science, see Sayer, 1992.)

Much the same occurs in human geography because of the nature of economic, social and political organisation. To survive in their physical environments, societies have developed means of organising themselves. A vast number of ways has been tried, but one now predominates throughout the world—capitalism. This way of organising life (or mode of production) is based on certain fundamentals, such as private property and profitability as the basis for production. Its particular dynamic contains a number of internal contradictions that have to be legitimated in order to ensure that its periodic crises do not lead to collapse (hence the need for a political organisation). That continuing dynamic, and the resolution of its crisis periods, has geographic components (as demonstrated by Harvey, 1982). The geography of capitalism has certain necessary features—most notably a pattern of uneven development in space and time—but their contours cannot be predicted. Which places prosper more than others at particular times is a result of the interaction of the general processes with the contingent circumstances of a myriad of places.

Places in this context are thus the arena within which the general laws of a global mode of production are interpreted and implemented by the local residents (sometimes successfully, sometimes not), in the light of their accumulated knowledge—their culture. Within those places, some people will probably be much more powerful than others in determining the local response, but learning from the experience of others (now much easier than ever before because of modern communications) means that frequently very similar programmes are launched in different places, though in particular ways because of the local political inheritance (as in the recent 'reconstruction' of the welfare state: Johnston, 1993). For example, in the UK the impact of greater market freedoms under Thatcherism led many local governments to adopt more 'entrepreneurial' approaches to local economic development. Some stood aside from this, promoting a more traditional, socialist approach in defiance of the general culture of the times: most soon failed, and eventually new leaders were elected who adopted the wider approach (as in the demise of Sheffield's 'Socialist Republic or South Yorkshire': Seyd, 1990). In many circumstances, however, what happens in a place is not a direct consequence of what its residents do, but rather of external perceptions of its nature: outside investors may withdraw their money from a place because they do not like certain aspects of its changing culture (as happened when Jamaica elected a slightly left-of-centre government in the late 1970s), and they may insist on changes before channelling investment to a place (as the IMF did to Jamaica in 1981, imposing conditions for support in regenerating the economy: the same happened in the UK in 1976).

The human geography of the world comprises a complex mosaic of places, therefore, each of which has a culture that has evolved as the residents continually re-evaluate their struggle for survival and assess the information that they receive from other places. A place's culture may change very substantially over a short period as its residents respond to particular pressures that threaten some aspect of their mode of living; elsewhere and elsewhen, change may be very much slower. So places are the milieux within which life is organised, and the units assessed by those who seek to change the world's economic, political and social geography—usually to serve their own ends.

What is a place?

Places vary substantially in their scale. They also overlap and intertwine in complex ways. Thus no one map of world places showing definitive boundaries can be drawn.

This is illustrated by the American South, which can be

clearly demarcated as the states which sought to secede
from the USA in the mid-nineteenth century over external
attitudes to a salient feature of their culture—slavery. The
outsiders were victorious, and the US Constitution was
amended to outlaw slavery and ensure that all are treated
equally. But the South remained a singular cultural region,
as the local mores were modified to bring it in line with the
Constitutional requirements and yet allow the powerful
white population to maintain its economic, social and
political superiority. Those mores involved male as well as
white superiority—most of the southern states never ratified
the Constitutional Amendment which gave women the
vote, for example (Johnston, 1991b). Similarly, the southern
state governments were less liberal in their implementation
of the nascent twentieth-century US welfare state legislation
(such as Medicaid and Medicare) than were their northern
and western neighbours, and local attitudes to trade unions
were among the most resistant in the entire country. Politically,
both the state governments and their representation in the
US Congress were dominated by the Democratic Party.

All of these aspects of the culture of the American South
were central to the socialisation of generations of their
residents for over a century: they defined the cultural
inheritance of new Southerners. They were also foundations
for major changes in the mid-twentieth century, induced by
external evaluations of the southern economy and aided by
the power of southern Representatives and Senators in
manipulating the pork barrel mechanisms which brought
much federal aid to their constituencies. Industries in the
north were facing crises of profitability, generated by high
costs and strong trade unions, among other factors. Investors
were looking for new places to put their money, where the
risk of losses and the cost and relative militancy of labour
were both relatively low. The South was very attractive, and
became the country's boom region, known as the 'sunbelt'.

The American South eminently qualifies for classification
as a place, therefore. It has been homogeneous in important
aspects of its culture which have influenced its position in
the changing map of American capitalism's uneven development during the twentieth century (on which see Markusen,
1987). But it is far from homogeneous on other criteria, and it
comprises a mosaic of places within places—down to the
smallest settlements with their own characteristics (as
illustrated by Harvey's (1993a) essay using a small settlement

in North Carolina to exemplify an argument regarding social justice). And the region is changing too: north-to-south migration of large numbers of relatively affluent people has made political attitudes in Florida much more Republican, for example.

Many other examples can be quoted of differences between places, at a variety of scales. Massey's (1984) work on the spatial redistribution of British industry away from the large conurbations—which stimulated a whole programme of research under the generic title of the 'localities' programme—illustrates this in a British context. (See also the essays in Massey, 1994).

Three key features of local cultures comprise their identifying characteristics: the physical environment, the built environment, and the people. Within the last of these, three further salient traits can be distinguished: social relations in the workplace; social relations in civil society (home and community); and the political institutions in each. These have been used as the framework for appreciating variations among coalfields in Great Britain in voters' attitudes towards the 1984–85 NUM strike: without appreciation of cultural differences between those places, the progress of the strike and, almost certainly, its eventual collapse in the face of government intransigence cannot be understood (Johnston, 1991b). Such work illustrates that (Harvey, 1993b, p. 5)

> while too much can be made of the universal at the expense of understanding particularity, there is no sense in blindly cantering off into that opaque world of supposedly unfathomable differences in which geographers have so long wallowed. . . . An account of the role of place in social life should prove helpful in this regard.

Human geographers can reach general conclusions about the operation of processes within the capitalist mode of production, therefore, and thereby come to appreciate its driving forces. What they cannot do is use that appreciation to predict what will happen where, because that is the outcome of the (personal and collective) learning and interpretation undertaken by knowing individuals. We can understand why, and where and when—but only by looking at singular decisions made within general contexts.

Making places

The argument so far has stressed the crucial role of places in the socialisation and acculturation of individuals. It is perhaps not surprising, therefore, that this important element of cultural life has been appropriated by those within society who wish to dominate it in some way, by manoeuvring the map of places (at whatever scale) and manipulating the people within them. Place-making is thus an important political activity. The reasons for it are varied, but many involve, in some way or another, promotion of the vested economic interests of particular local groups. (Exceptions include non-economic cultural interests, such as those of Bosnia's and Rwanda's 'ethnic cleansers'.)

A defining characteristic of a place is that its location and extent are known; that is often used as a strategy for organising society, and in particular exercising power within it. *Territoriality* is a very effective means of controlling people, for example (Sack, 1983, 1986). A bounded place has clear territorial expression. Its boundaries can be identified—on the ground and on maps—they can be defended; and they can be communicated. To be within the boundaries of a place is to be subject (willingly or not) to the control of those with power there. Thus it is a ready form of classification; those subject to the sovereign power of a state are those within its territorial boundaries.

A territoriality strategy is also important because it enables its implementers to displace and depersonalise the notion of control. Within a state's borders, for example, you are subject to 'the law of the land', not the laws imposed by the powerful, which deflects attention from the inequalities inherent in all capitalist (and most other) societies. This operates at a variety of scales, and need not involve the power of the state. I define my office—my territory—as a no-smoking zone: notices indicate that is the rule which is effective there, not that it is my rule.

Territoriality is crucial for nation-states because it provides a definable container within which the state's power is sovereign (Gottmann, 1951, 1952; Taylor, 1994). That container is crucial to the state's fundamental role in sustaining the mode of production. Capitalism must be regulated, otherwise anarchy would prevail. Regulation—of money and credit, of contract compliance, of environmental pollution, etc.—is facilitated by, indeed may be impossible without, a territoriality strategy, hence its crucial role in capitalist state activities.

Appreciating place-creation by the state is crucial to understanding the operation of contemporary society, therefore. The capitalist space-economy is based on competition, though because there is a very fine distinction between competition and conflict, geopolitical tensions often result (as argued in another of Harvey's (1985) seminal essays). Promotion of the interests of (the people in) one place must be done very largely by opposing the interests of (those in) other places, which often involves promoting positive views of oneselves and negative views of others. This is frequently observed within the European Union, with some British politicians promoting negative images of the nationals of other states; an excellent example of the practice is the magazine *The Reader's Digest*, which for long cultivated a positive view of American values, goals and structures that was starkly contrasted with very negative opinions of the Soviet Union (Sharp, 1993).

Places are used as the containers within which to create residents' positive self-identities, therefore, and also, by default if not design, negative identities of the dwellers in other places. This is superbly illustrated in the contemporary world by nationalist movements, as groups with a clear self-identity but no exclusive territory are seeking one, whereas others with a territory are mobilising support for their claims of sovereignty there. The world is in continuous place-making flux, and the antagonisms over places all too often become armed conflicts. Indeed the spatial separatism of the antagonist groups can deepen the mistrust between them, and foment the conflict—as in the highly segregated city of Belfast, where for 25 years each side, relatively secure in its own ghetto, fostered a negative view of the other.

Place-making is important at the substate level also. Cultural identification with a place and some representation of it—such as a sports team—is a common means of mobilising identity. Bringing together the various interests in a place is also significant in the competition for economic success which is becoming increasingly tough. Place-based alliances promote local economic development within the UK, for example, whether based on local governments, Chambers of Commerce, or Training and Enterprise Councils.

Learning about places

Places are crucial to the study of both human and physical geography, for similar and different reasons. I have concentrated here on their importance in human geography, and turn now to how that importance can be recognised within the educational process.

The nature of places can be summarised in the following points (drawing on Johnston, 1990).

1. *Places are social creations*: they have separate identities because people have made them so. Those creations may be based in the local physical environment, because different 'natural' milieux may stimulate different human responses (although the same milieux may stimulate different responses in different people—or even, perhaps, in the same people at different times). As in so much of what we study, the environment is both enabling and constraining: it offers us choices but simultaneously limits what we might do.

2. *Places are self-reproducing*, because people are socialised into the local culture and, by their actions thereafter, participate in its re-creation. People are made in places, so if places differ people will too, though not in a deterministic way. In turn, they will ensure that places continue to differ.

3. *People control and change places*. A place-bound culture has no identity separate from its current members, and they may (deliberately or by default) change that culture because of the need to respond to new stimuli or to make contact with people in other places (which increasingly may be indirect, through mass media). The determined actions of powerful local individuals, who wish to steer the place in a particular direction for their own ends, may also result in change. Furthermore, over time certain aspects of a culture may be discarded, or at least forgotten. What will be discarded, what retained, and what adopted (and perhaps adapted) from outside cannot be predicted: it depends on the people themselves, because culture (like language) both constrains and enables.

4. *Places are not isolated and self-contained*. Globalisation is a contemporary 'keyword', reflecting and increasing integration—economically, socially, politically and culturally—of its disparate parts (Johnston and Taylor, 1989; Johnston et al., 1995). The earth has always been a single interdependent physical system; increasingly the same description applies to its human organisation. People, and hence places, are

subject to the globalising forces of the world economy. They interpret those forces in the context of their own cultural backgrounds, and respond accordingly, thereby contributing both to their place's evolution and to the ever-changing map of the global society.

5. *Places are often formally bounded*, in order to promote their separateness and the jurisdiction of those with power there (whether taken or granted). Such formally bounded places are used to promote the inhabitants' sense of identity, counterposed to a sense of difference from those who live elsewhere. This frequently involves stereotyping.

6. *Places are potential sources of conflict*, and wars are almost invariably fought between places—between states which claim a monopoly of legalised violence within modern society.

The educational task

Understanding places is fundamental to understanding the world, therefore. But how do we do it? Geographers have strayed down a number of culs-de-sac in their search for a disciplinary focus in recent decades. One of those was regional geography, a now largely discredited operation which provided descriptive material about places, frequently in little more than catalogue form and often with no explanatory structure—apart from a thinly disguised environmental determinism in a number of cases. My arguments here about the importance of place are not leading to a call for a revival of such regional geography, in its original or any other form.

The desiderata for geographical education in this context are that:

1. it stresses the differences between places, as separate milieux; and yet
2. it also stresses the importance of places as contexts within which general processes—whether human or physical—are played out.

Because places differ, so the operation of the processes differs too. To this end

3. it makes the role of people in making places clear, as cultural constructions which are continually being reproduced and changed.

As a consequence

4. geographical education should enable students not only to appreciate their own places (and personal 'sense of place') but also to draw on that appreciation in developing their understanding of the wider world.

In undertaking those tasks, geographical education involves both describing the differences between places and yet using those differences to identify the operation of general processes. To do one without the other is to mislead: to stress differences is to be overly individualistic; to stress general processes is to be deterministic.

The task is difficult yet immensely challenging and potentially satisfying. It stimulates understanding of the great variety of places which comprise the earth; it excites interest in the general forces which underpin the operation of both physical and human environments; it arouses appreciation of the role of human creativity in the construction of the contemporary mosaic of places; and it ensures humility in realisation of the constraints to that creativity. But perhaps more important than anything else, it aids the creation of an informed world citizenry, whose beliefs are based not on stereotypes but on appreciation of, and willingness to work within, a world of difference and variety.

Much that Karl Marx said has been challenged, and many have dismissed it all. One of his most-cited generalisations is that we make our own histories, but not in circumstances of our own choosing—and on this, if nothing else, he got it exactly right. Places are those circumstances and our task as geographers is to understand them, and so promote appreciation of how our histories have been, are being, and will be made.

References

Berry, B.J.L. (1964) Approaches to regional analysis: a synthesis. *Annals of the Association of American Geographers*, **54**, 2–11.
Curry, L. (1962) Climatic change as a random series. *Annals of the Association of American Geographers*, **52**, 21–31.
Gottmann, J. (1951) Geography in international relations. *World Politics*, **3**, 153–173.

Gottmann, J. (1952) The political partitioning of our world: an attempt at analysis. *World Politics*, **4**, 512–519.

Gould, P.R. (1979) Geography 1957–1977: the Augean period. *Annals of the Association of American Geographers*, **69**, 139–151.

Gould, P.R. (1985) *The Geographer at Work*. Routledge, London.

Harvey, D. (1982) *The Limits to Capital*. Blackwell Publishers, Oxford.

Harvey, D. (1985) The geopolitics of capitalism. In D. Gregory and J. Urry (eds), *Social Relations and Spatial Structures*, Macmillan, London, pp. 128–163.

Harvey, D. (1993a) Class relations, social justice and the politics of difference. In M. Keith and S. Pile (eds), *Place and the Politics of Identity*, Routledge, London, pp. 41–66.

Harvey, D. (1993b) From space to place and back again: reflections on the condition of postmodernity. In J. Bird et al. (eds), *Mapping the Futures: Local Cultures, Social Change*, Routledge, London, pp. 3–29.

Johnston, R.J. (1989a) *Environmental Problems: Nature, Economy and the State*. Belhaven Press, London.

Johnston, R.J. (1989b) Philosophy, ideology and geography. In D. Gregory and R. Walford (eds), *Horizons in Human Geography*, Macmillan, London, pp. 48–66.

Johnston, R.J. (1990) The challenge for regional geography: some proposals for research frontiers. In R.J. Johnston, J. Hauer and G.A. Hoekveld (eds), *Regional Geography: Current Developments and Future Prospects*, Routledge, London, pp. 122–139.

Johnston, R.J. (1991a) *Geography and Geographers: Anglo-American Human Geography since 1945* (4th edn). Edward Arnold, London.

Johnston, R.J. (1991b) *A Question of Place: Exploring the Practice of Human Geography*. Blackwell Publishers, Oxford.

Johnston, R.J. (1993) The rise and decline of the corporate welfare state: a comparative analysis in global context. In P.J. Taylor (ed.), *Political Geography of the Twentieth Century: A Global Analysis*. Belhaven Press, London, pp. 115–170.

Johnston, R.J. and Taylor, P.J. (eds) (1989) *A World in Crisis? Geographical Perspectives*. Blackwell Publishers, Oxford.

Johnston, R.J., Taylor, P.J. and Watts, M.J. (eds) (1995) *Geographies of Global Change: Critical Perspectives*. Blackwell Publishers, Oxford.

Livingstone, D.N. (1992) *The Geographical Tradition: Episodes in the History of a Contested Enterprise*. Blackwell Publishers, Oxford.

Markusen, A. (1987) *Regions: the Economics and Politics of Territory*. Rowman and Littlefield, Totowa, NJ.

Massey, D. (1984) *Spatial Divisions of Labour*. Macmillan, London.

Massey, D. (1994) *Space, Place and Gender*. Polity Press, Oxford.

Robinson, B.S. (1973) Elizabethan society and its named places. *The Geographical Review*, **63**, 322–333.

Sack, R.D. (1983) Human territoriality: a theory. *Annals of the Association of American Geographers*, **73**, 55–74.

Sack, R.D. (1986) *Human Territoriality: Its Theory and History*. The University Press, Cambridge.

Sayer, A. (1992) *Method in Social Science* (2nd edn). Routledge, London.

Seyd, P. (1990) Radical Sheffield: from socialism to entrepreneurialism. *Political Studies*, **38**, 335–344.

Sharp, J.P. (1993) Publishing American identity: popular geopolitics, myth and *The Reader's Digest. Political Geography*, **12**, 491–504.

Sitwell, O.F.G. (1993) *Four Centuries of Special Geography*. UBC Press, Vancouver.

Taylor, P.J. (1994) The state as container: territoriality in the modern world-system. *Progress in Human Geography*, **18**, 151–162.

Only connect: approaches to human geography

Peter Jackson

Introduction

This chapter aims to provide a critical perspective on the wide range of (often competing) approaches to human geography that have emerged over the past 10 or 20 years. It is, to some extent, a personal reading of the recent history of human geography, written from the perspective of a sometimes sceptical but still enthusiastic practitioner. While some have lamented the demise of a unified geography, deploring the growing gap between human and physical geography and the proliferation of ever more inward-looking specialisms (Stoddart, 1986), this chapter takes a different stance, celebrating the plurality and vibrancy of contemporary human geography and searching out possible connections between different styles of research at different scales of analysis. It also attempts to draw out the implications of some of these recent changes for the teaching of geography in schools. I begin, though, by outlining what can be regarded as the conventional wisdom about the philosophy of human geography and questioning the alleged fragmentation of the subject.

Received wisdom?

Since the 1980s, approaches to human geography have frequently been summarised in terms of *three main philosophical*

Geography into the Twenty-first Century, Edited by E.M. Rawling and R.A. Daugherty,
© 1996 John Wiley & Sons Ltd.

approaches: positivism, humanism and structuralism. Positivism can be briefly characterised as the (often unacknowledged) philosophy that lay behind geography's 'quantitative revolution' in the 1960s and which underpinned the development of locational analysis, statistical modelling and other forms of spatial analysis. The second philosophical approach (humanism) was highly critical of all such claims to a universal and dehumanised 'scientific' geography, emphasising instead the role of personal subjectivity, individual creativity and the scope of human agency. Inspired by the philosophies of existentialism, phenomenology and idealism, geographical humanism flourished in the 1970s, first as a modified form of positivism (behavioural geography) and then as a more full-blown critique of scientism (humanistic geography). The third philosophical perspective (structuralism) was associated with the rise of social unrest throughout Europe and North America in the 1960s and led to calls for an explicitly 'radical' geography. Though Marxism cannot be equated unproblematically with structuralism (some versions of Marxism being more akin to humanism in their emphasis on class struggle and human agency), it was a particular brand of structural Marxism (associated with the philosopher Louis Althusser and the sociologist Manuel Castells) which proved particularly attractive in the search for structural explanations of historical and geographical change (and which later proved an easy target for their critics). The emphasis of most structuralist accounts was on the economic basis of class exploitation. Other forms of oppression and social stratification, articulated through 'race', gender or sexuality, for example, were not so well developed at this time.

Though far from exhaustive, this highly simplified account of recent intellectual history has proven a popular and influential way of thinking about the philosophy of human geography. The tripartite structure of positivism, humanism and structuralism was clearly discernible in Ron Johnston's best-selling *Geography and Geographers* (1979 and subsequent editions). It was the organising framework of his *Philosophy and Human Geography* (1983) and was also chosen as the basis for Jim Bird's *The Changing Worlds of Geography* (1989). The framework persists (with the inclusion of two additional 'approaches': realism and postmodernism) in Cloke et al.'s *Approaching Human Geography* (1991). It was also employed in Tim Unwin's *The Place of Geography* (1992) where it was interpreted in relation to Habermas's argument about three contrasting forms of knowledge (empirical–analytic, historical–hermeneutic and critical), linked to their respective material interests (in technical control, self-understanding and human emancipation).

Students find these accounts appealing as a way of organising the diversity of knowledge and approach with which they are confronted. Such frameworks allow them to contrast perspectives, to criticise different approaches and to 'pigeon-hole' ideas. But they tend to reduce the complexity of geographical knowledge to various kinds of *linear narrative* (a heroic reading of history, where 'progress' is inevitable as positivism overcomes the descriptive tendencies of an earlier regional geography only for spatial analysis to be superseded by humanist and radical critiques). Students also tend to use such frameworks as a way of labelling authors and ideas rather than focusing on the *tensions between different perspectives*: subjective/objective, fact/value, structure/agency, etc. They are reluctant to explore ambiguity and shifting allegiances, wanting to pin things down definitively, once and for all. Similar debates about the over-simplification of ideas and approaches to human geography were played out in the development of the Geography 16–19 Project (Naish and Rawling, 1990; Naish et al., 1987). Following Johnston (1983), the project recognised three philosophies that were currently dominating current work in human geography: positivist, humanistic and structuralist. But rather than arguing that the curriculum should be based on one approach or paradigm, the project team decided to start from a range of questions, issues and problems and to involve students in selecting ideas and methods from across the range of academic approaches to the subject. (For a discussion of the limits of such philosophical eclecticism, see Eyles and Lee, 1982.) Despite these reservations, the 'philosophical triad' remains a useful starting-point for thinking about the history of geographic thought and the complex structure of geographic knowledge.

Fragmentation?

The 1980s were also often seen as a period of *fragmentation* when geography's 'core' interests (in regional differentiation, landscape evolution and human–land interaction, for example) were replaced by a proliferation of subdisciplinary themes. While some new themes ('sustainability' is the most obvious) are integrative, combining human and physical geography, there were also divisive tendencies such as the proliferation of new terminology, much of it borrowed from the social sciences and the humanities, that has tended to drive a wedge between the social and environmental branches of the discipline. This tendency can be gauged by comparing successive editions of *The Dictionary of Human Geography* (Johnston et al., 1994) which has expanded from 500 to 700 entries and from 18 to 45 contributors since the first edition in 1981. The process can also be traced through the profusion of study groups within the Institute of British Geographers (IBG) and via the burgeoning 'specialty groups' of the Association of American Geographers (Dear, 1988). It can also be seen in the emergence of new journals which cater to 'special interests' (in cultural, feminist or regional geography, for example) rather than to the 'common ground' of geography's allegedly core interests.

Geography's supposed fragmentation has been deplored by many of the discipline's stalwarts such as David Stoddart (1987) who urges us to abandon excessively narrow or trivial interests and to (re)claim the high ground. His examples of the former distractions are (apparently fictitious) projects on 'geographic influences on the Canadian cinema' and 'the distribution of fast-food outlets in Tel Aviv'. His example of 'reclaiming the high ground' is the kind of study that he might himself undertake on the relationship between population, resources and environmental hazards in coastal Bangladesh. But there is another side to his argument. Those who claim the 'high ground' for their own kind of work inevitably (if sometimes unintentionally) marginalise other people's interests. Stoddart's subsequent rebuttal of a feminist critique of his reading of the discipline's history indicates the less benign aspect of the definer's art (Domosh, 1991; Stoddart, 1991). For any attempt to define a 'core' (whether it be in terms of landscape change, environmental management or human–nature interactions) inevitably leads to the exclusion or peripheralisation of other kinds of geography, especially those with a less-established claim to inclusion on the geographical agenda. What looks like fragmentation from the 'centre' may seem entirely different from the 'margins'. From a marginalised perspective, fragmentation may look like a welcome plurality, offering a range of alternative spaces in which to articulate sometimes unpopular views.

Geographers have begun to theorise these ideas in terms of the concept of *positionality* (Jackson, 1993). They suggest not only that there are multiple 'ways of seeing' the world but that each perspective is partial (in both senses of the term: incomplete and biased). We should therefore be suspicious of anyone who claims the superiority of their own position or who implies that their particular 'way of seeing' is uniquely privileged. Significantly, it is privileged perspectives that most often represent themselves as neutral and marginalised voices that are most often criticised as biased. Consider the all-too-frequent accusation that feminism is politically biased, overly emotional or otherwise partial. Masculinist histories are rarely treated in the same way (though see Rose's (1993) critique of the epistemological biases of 'mainstream' geography). Reversing the traditional line of argument, it is possible to claim that the adoption of a particular (feminist) standpoint is a source of strength not weakness, and to insist that it is those who do not admit their partiality whose perspective is likely to be most seriously flawed. Whether similar arguments can be applied to other unexamined discourses within human geography (such as those associated with whiteness or masculinity or heterosexuality) remains to be seen, though the early indications are certainly promising (e.g. Valentine, 1993; Pile, 1994).

Common ground?

In the face of so much discussion of the discipline's supposed fragmentation it was perhaps inevitable that some would strive for a renewed unity. One such argument was articulated by Ron Cooke in his presidential address to the IBG. Here, Cooke sought to map out a disciplinary sense of 'common ground, shared inheritance', coupled with a series of 'research imperatives' for environmental geography (Cooke, 1992). The implied centralism (and coercive tone of the 'imperatives') can be countered by arguing that many of the most exciting recent developments (in the social sciences at least) have emerged *at the margins* of established disciplines. Indeed, marginality has become one of geography's keywords in the 1990s, explored at length (by a sociologist) in Rob Shields's *Places on the Margin* (1991). Developments in feminism, cultural studies and post-colonial criticism are just three of

the most conspicuous recent interdisciplinary examples, all
of which are having a significant impact on human geography
(Rose, 1993; Jackson, 1989; Radcliffe, 1994). Now these may
be exactly the kinds of issues that former Secretary of State
Kenneth Baker had in mind when he referred to 'vague
concepts and attitudes which relate to various subjects' as
opposed to learning about real places and where they are
(*Hansard*, 29 April 1991). But we ignore such developments
at our peril. Taken seriously, they do not simply add new
dimensions to our existing research agenda. They have the
potential to *transform* those agendas, reshaping the way we
view the world (as feminism did with economic geography,
redefining our concepts of skill and work, redrawing the
boundaries between 'public' and 'private', 'home' and
'work'). As educators, we need to think carefully about how
to translate these ideas into the curriculum, conscious of the
ever-shifting dividing lines between politics and pedagogy
(Giroux and McLaren, 1994).

A new pluralism?

Reflecting these issues, recent histories of geographical
thought have abandoned a single, linear narrative leading
triumphantly to the present. This new pluralism is evident
in David Livingstone's richly rewarding study of *The
Geographical Tradition* (1992), whose subtitle ('episodes in the
history of a contested enterprise') belies the implied singularity
of his title. In the frontispiece to that book, Livingstone
quotes Clarence Glacken's remark that 'a historian of
geographical ideas . . . who stays within the limits of his [*sic*]
discipline sips a thin gruel' with obvious approval. His own
telling of geography's story ranges widely across the history
and philosophy of science, encompassing the various contexts
in which geographical ideas (from travellers, explorers and
academics) were received and interpreted by the population
at large.

A similar pluralism is also apparent in Derek Gregory's
last-minute decision to retitle his study of 'the geographical
imagination' in the plural (*Geographical Imaginations*, 1994).
Gregory draws on a wide range of social and literary theory,
remapping geography into the human sciences, revealing
the limitations of the discipline's Eurocentrism, highlighting
its preoccupation with the visual, and making connections

between various geographical discourses and those of multiple other histories. In each case, the encounter between geography and other branches of the human sciences reveals a world of multiple and contested meanings, raising exciting challenges for geographical research and teaching. In the process, traditional interests in landscape, space and place are transformed by their encounter with social theory. A healthy suspicion of disciplinary boundaries is entirely characteristic of Gregory's (1994, p. 51) attempt to recover a less inhibited geographical imagination:

> [A] sense of separateness was to become the enduring legacy of the fin-de-siècle encounter between geography and sociology . . . frozen into disciplinary boundaries, they were patrolled and policed; and once identity papers had to be produced for inspection, travellers were threatened with exile or internment: there was precious little space to question those differences, still less to transcend them.

However expedient the insistence on the distinctiveness of a geographical perspective, whether defined in terms of content or method, it is the liberating sense of transcendence and intellectual freedom that are threatened by narrowly defined disciplinary prerogatives. It should not be surprising, therefore, that the kind of theoretical developments discussed in this chapter have been accompanied by a refreshing willingness to experiment with a wide range of research methods from ethnography (first-hand field research and textual (de)construction) to iconography (the interpretation of culturally significant signs and symbols). If we are witnessing a return to 'regional geography' (as many have suggested), it is mainly distinguished from its predecessors by the reflexivity of its method and by a welcome self-consciousness about narrative strategy (Sayer, 1989).

Beyond the fragments?

For those who lament geography's supposed fragmentation, the blame is often laid at the door of postmodernism, one of the most complex and contested terms in our contemporary intellectual vocabulary. Variously defined as a self-consciousness about language and culture (referred to as 'dual coding' by Charles Jencks) and as a rejection of grand theory (or 'meta-narrative'), others have concluded cynically that 'This word has no meaning. Use it as often as possible' (*The Independent*, 24 December 1987). While some argue that postmodernism represents a dramatic break from modernism, others emphasise the continuities and refer to hyper- or late-modernism. Underlying many debates has been the question of whether aesthetic shifts in taste and style can be causally linked to economic changes towards a post-Fordist regime of flexible accumulation. For some observers, post-modernism is little more than 'the cultural logic of late capitalism' (Jameson, 1984).

Among its proponents, postmodernism is celebrated as the dawn of 'new times' where new spaces are being opened up for formerly marginalised groups to express their views. Older, class-based and masculinist versions of politics are being challenged by new political movements around gender and sexuality, the 'new ethnicities', the politics of the body (reproductive rights, health care and disability), and pro-gressive forms of regionalism and nationalism. Among its critics, however, postmodernism is associated with the descent into a moral vacuum, a self-indulgent and formless relativism where the old certainties no longer hold and 'anything goes'. Responding to this criticism, geography's most public engagements with postmodernism have attemp-ted to avoid such relativism and to retain an overall map by which to make sense of the world. David Harvey's analysis of *The Condition of Postmodernity* (1989) is perhaps the most striking example, charting the transition from modernism to postmodernism (in the cultural sphere) and the parallel transition from Fordism to post-Fordism (in the economic sphere). Harvey links these simultaneous transitions via the concept of '*time–space compression*', which seeks to account for a shrinking world and a destabilised sense of time through an analysis of the social impact of technological change and the ever-shortening time of capital circulation. Harvey has been criticised for his linear histories and dichotomous logic (modernism/postmodernism, Fordism/post-Fordism)—themselves highly modernist ways of think-ing. The problems of constructing any such map of the contemporary world that does not somehow position its author in an impossible space outside of that world have also been identified. Problems of representation and partiality have also been raised by feminist critics (e.g. Deutsche, 1991; Massey, 1991a).

In *Postmodern Geographies* (1989), Ed Soja attempts to construct an alternative to the traditional linear narrative, taking Los Angeles apart and recombining the fragments. It is a more playful reading of postmodernism than Harvey's. There is, Soja tells us, 'a Boston in Los Angeles, a Lower Manhattan and a South Bronx, a São Paulo and a Singapore' (p. 193), combining the seemingly paradoxical but *functionally interdependent* worlds of the 'haves' and the 'have-nots'. While he ends up with a very conventional geography of the city—a kind of contemporary version of Burgess's concentric zone model: the industrial city turned inside out—Soja establishes the crucial need to connect the apparently disconnected, that is such a central feature of contemporary human geography.

The global and the local?

Much of the current theoretical literature in human geography is about *making connections*: between the apparent dualisms of land and life, people and place, society and space. The National Curriculum encourages geography teachers to focus on particular places and on a variety of scales from the global to the local. But how are we to make the connections between these different scales? All too often, at all levels of geographical education, we present students with case studies at a variety of different scales, leaving them to do the imaginative work of connecting the various scales. At the conference from which these chapters were drawn, Margaret Roberts suggested a photographic analogy to me after my talk, starting with a close-up of the local area and panning out to the regional, national or global scale, or starting globally and zooming in the micro-scale. It is these connections that I wish to explore in the remainder of this chapter. I shall argue that the kind of diversity that now characterises geography in higher education could be extended to the teaching and learning of geography in schools. A respect for diversity and an exploration of difference are fundamental characteristics of the geographical imagination and should not be forced off the agenda by the centralising tendencies of standardisation and uniformity.

Since the source of my title ('only connect') and the inspiration for this chapter comes from the epigraph to E.M. Forster's novel *Howards End* (1910), it seems appropriate to

begin with a literary example. The following extract from
Jamaica Kincaid's extended essay on her native Antigua
provides many such creative connections. Here, she mimics
the tourist's innocent gaze only to undermine it with some
more down-to-earth connections (Kincaid, 1988, pp. 12–14):

> [Arriving in Antigua, you] long to refresh yourself; you
> long to eat some nice lobster, some nice local food. You
> take a bath, you brush your teeth. You get dressed
> again; as you get dressed, you look out the window.
> That water—have you ever seen anything like it? Far
> out, to the horizon, the colour of the water is navy-blue;
> nearer, the water is the colour of the North American
> sky. From there to the shore, the water is pale, silvery,
> clear, so clear that you can see its pinkish-white sand
> bottom. ...
>
> You see yourself taking a walk on that beach, you see
> yourself meeting new people. ... You see yourself, you
> see yourself. ... But you must not wonder what exactly
> happened to the contents of your lavatory when you
> flushed it. You must not wonder where your bath-water
> went when you pulled out the stopper. You must not
> wonder what happened when you brushed your teeth.
> Oh, it might all end up in the water you are thinking of
> taking a swim in; the contents of your lavatory might,
> just might, graze gently against your ankle as you wade
> carefree in the water, for you see, in Antigua, there is no
> proper sewage-disposal system. But the Caribbean Sea
> is very big and the Atlantic Ocean is even bigger; it
> would amaze even you to know the number of black
> slaves this ocean has swallowed up. ... There is a world
> of something in this, but I can't go into it right now.

The need to make connections between geographical scales
and across historical time has started to preoccupy some of
the discipline's most innovative writers. Here, I shall refer to
just three: David Harvey, Neil Smith and Doreen Massey,
before concluding with an example from the world of
curriculum development at school level.

David Harvey (1990, p. 422) is fond of telling the following
story which is designed to encourage his students to make
connections between the local and the global:

> I often ask beginning geography students to consider
> where their last meal came from. Tracing back all the
> items used in the production of that meal reveals a

relation of dependence upon a whole world of social labor conducted in many different places under very different social relations and conditions of production. That dependency expands even further when we consider the materials and goods used in the production of goods we directly consume. Yet we can in practice consume our meal without the slightest knowledge of the intricate geography of production and the myriad social relationships embedded in the system that puts it upon our table.

Similar arguments can be made for many other examples where nature has been refashioned for human use (Goodman and Redclift, 1991), underlining the fact that the environment is socially constructed and that resources are always subject to cultural appraisal.

A series of connections across scales and over time is also made in Neil Smith's recent work on neighbourhood change and gentrification in Lower Manhattan, symbolised in the conflict over Tompkins Square Park on the Lower East Side. Unknown to most of the participants in the Tompkins Square riots of August 1988, Smith uncovers a series of parallels with the riots of 1874 which also raised questions of ownership, access and the control of public space. Smith (1993, p. 91) builds outwards from this specific example to a general argument about 'scaling places':

Homelessness and gentrification . . . defined both the text and the subtext of the police riot that took place in and around Tompkins Square Park in the Lower East Side. ... Gentrification began pulsing through the area from west to east in the late 1970s and accelerated in the 1980s. Its causes were global as much as local: the rapid expansion of the world financial markets focused on Wall Street and the adjacent Financial District; national economic expansion following the recessions of 1973 to 1982; recovery from the city's fiscal crisis; the availability of a dramatically undervalued stock of tenement buildings, resulting from decades of disinvestment intensified since the 1950s; the planned as well as spontaneous centring of an alternative art industry in the area which became the cultural anchor around which reinvestment hype could be organized; and the active encouragement of myriad city and state pro-grammes devoted to housing rehabilitation and re-development, anti-drug and anti-crime campaigns, and a park reconstruction programme.

Note, too, that the connections Smith identifies are political as well as intellectual, defining the scale at which the contest for space was being fought out. Smith (1993, p. 93) demonstrates the escalating scale of the riot, from the Square itself to a broader city-wide struggle:

> In the first days after the riot, there was an explosion of graffiti in the neighbourhood around the park, directly commenting on the riot, gentrification, displacement, the financial crash of the year before, and the social purposes of art. Stencil artists specifically retaliated against the official definition of the park and its residents, and the implied confinement of struggle, by scripting the park's entire vicinity as an 'NYPD RIOT ZONE' [New York Police Department]. At the same time, hastily constructed links were built between park residents, squatters and housing activists and 'Whose park is it?' was replaced by the slogan: 'Tompkins Square Everywhere!'

Recognising the political potential of such spatial metaphors, Smith concludes that we need to develop a more effective language of spatial difference and differentiation. His own analysis of the production of scale works outwards from the body through the home and the community to the urban and regional, ultimately to the national and the global. But 'the active social connectedness of scales' of which he speaks remains elusive (1993, p. 101). It is one thing to invoke Cynthia Enloe's aphorism that 'The personal is international' (1990, p. 195), another to *demonstrate* that inter-connectedness.

Recent work has also begun to make the connections between a localised sense of place and a sense of global belonging. While some readings of postmodernism imply that a sense of place can only be understood as a *retreat* to a nostalgic past world, a lament for an unmediated lost 'authenticity', there are more congenial alternatives such as Doreen Massey's (1991b) arguments for a progressive 'global sense of place'. Rather than seeing people's attachments to place as a backward-looking yearning for lost identity or a threatened cultural heritage, Massey insists on the potential for seeing places as a unique combination of processes working at a variety of scales: a particular moment in a wider web of intersecting social relations. Rather than trying to draw boundaries around places, she talks about open and porous networks of social relations. The unique identity of any particular place is constructed through the specificity of its associations with other places. Taking the example of her local high street in Kilburn she attempts to define a progressive sense of place that is less about fixity and the singularity of an identity *rooted* in the past, and more about the dynamism and plurality of identities that are defined in terms of people's multiple *routes* through the place where they live (Massey, 1994, pp. 152–153):

> [Kilburn] is a pretty ordinary place, north-west of the centre of London. Under the railway bridge the news-paper stand sells papers from every county of what my neighbours, many of whom come from there, still often call the Irish Free State. The postboxes down the High Road, and many an empty space on the wall, are adorned with the letters IRA. ...
>
> Thread your way through the often almost stationary traffic diagonally across the road from the newsstand and there's a shop which as long as I can remember has displayed saris in the window. ... Overhead there is always at least one aeroplane—we seem to be on a flight-path to Heathrow and by the time they're over Kilburn you can see them clearly enough to tell the airline and wonder as you struggle with your shopping where they're coming from. Below, the reason the traffic is snarled up... is in part because this is one of the main entrances to and escape routes from London, the road to Staples Corner and the beginning of the M1 to 'the North'. ...
>
> . . . while Kilburn may have a character of its own, it is absolutely not a seamless, coherent identity, a single sense of place which everyone shares. It could hardly be less so.

From this perspective, Massey concludes, it is impossible to think about Kilburn High Road without bringing into play half the world and a considerable amount of British imperialist history too (1994, p. 154). The importance of such a global sense of place, Massey insists, is that it offers a sense of local identity that is open-ended, extroverted and socially inclusive. By being conscious of our links to the wider world, we are less inclined to fall back on narrowly defined, introverted or socially exclusive forms of place-identity, the kind that bedevil the contemporary world from the level of international politics to the 'nationalism of the neighbourhood'.

There are still too few examples of this kind of work in the educational literature, though the work of the London-based

Association for Curriculum Development and the Birmingham Development Education Centre are important exceptions. In their adoption of a 'development education' approach to geography at Key Stage 3, for example, the Birmingham Centre argue that adopting a global perspective implies not only appreciating one's own position in the world but also an exploration of what is of value in other ways of viewing the world. They advocate 'starting local', with issues such as the closure of a local factory, on the basis that students are likely to be familiar with their local area or home region, before exploring the 'global dimension' in the local area (Development Education Centre, 1992).

A brilliant example of how to combine the local and the global was produced by the Association for Curriculum Development. *Sweet or Sour?* is a book-length project on the Jamaican sugar industry (Larkin and Widdowson, 1988). Rather than studying Jamaica as some far-off, exotic island, the authors traced the sugar industry's multiple links back to Britain through the history of slavery and imperialism up to the present-day operations of transnational corporations such as Tate & Lyle. Using a series of ingenious and imaginative exercises, readers were introduced to the history of British involvement in the Caribbean, the politics of industrialised agriculture in Jamaica and the social consequences of plant closures in the East End of London (where Tate & Lyle had a factory). While the project made no bones about its own political agenda, the material provided was open-ended and flexible, capable of different interpretations and striving to make connections with students' own personal experience. (Examples included Caribbean products that are widely available in the UK, a letter-writing exercise to an imaginary Jamaican pen-pal, and references to popular Jamaican music.) Such examples suggest the enormous potential for further creative interventions, emphasising the connectedness of our global economy while providing a richness of local detail on particular places.

Summary and conclusion

This chapter began with an all-too-hasty survey of the contested philosophical bases of contemporary human geography. It argued that what many observers have seen as a regrettable fragmentation of the subject, between human and physical geography and within particular specialisms, could be seen more positively as a welcome plurality of method and approach. Following an endorsement of the benefits of interdisciplinarity, a 'politics of position' was sketched which emphasised the dangers of defining a rigid disciplinary core, consigning others to the margins, and challenged the apparent neutrality of various dominant discourses. Writing 'from the margins' and from an avowedly committed standpoint may offer a perspective which is preferable (if self-consciously partial) to the unexamined biases of the privileged 'centre'. New ways of writing the history of geography that reject linear narratives and attempt to accommodate a plurality of subject positions were reviewed, leading to an evaluation of geography's often troubled encounter with postmodernism. Finally, the importance of making connections across time periods and between places at a variety of geographical scales was emphasised, with some examples from the recent academic and educational literature.

An emphasis on theoretical pluralism and an exploration of the connectedness between scales are by no means a comprehensive agenda for human geography in the twenty-first century. But they are important components of such an agenda. Plurality is not to be welcomed for its own sake but because of the political choices that such an approach highlights. The 'celebration' of intellectual diversity is ultimately a barren approach to the history of ideas if it is not simultaneously coupled with an emphasis on the politics of knowledge. It is up to us, as teachers, to find the most stimulating ways of conveying the idea that there are multiple ways of seeing the world, that all forms of knowledge are selective, and that the partiality of particular forms of knowledge is indicative of a politics of position. As Stuart Hall has argued, 'We all speak from a particular place, out of a particular history, out of a particular experience, a particular culture, without being contained by that position' (1992, p. 257).

Above all, I have argued, we need to help our students see the connections between people's 'imagined communities' (Anderson, 1983) at a variety of scales from the global to the local, from the nation to the neighbourhood. It is this articulation between scales that represents one of our most significant intellectual challenges in the 1990s, not whether some fossilised 'core' of a unified geography is being eroded from the margins. While the politics of the recent past may have forced us into an increasingly entrenched position, it may now be time to look outwards again. Rather than policing our own disciplinary boundaries in an embattled and defensive mood, I would suggest that it is human geography's encounter with social theory and its excursions into neighbouring social sciences that are the most promising sources for meeting the intellectual and political challenges of the future. This poses significant challenges for geographical education in the twenty-first century and underlines the need for teachers in higher education and at school to continue the dialogue that was engaged at the Oxford conference in 1994.

Acknowledgements

Thanks to my colleagues Nicky Gregson and Andy Wood, and to the editors, for their constructive comments on earlier drafts of this chapter.

References

Anderson, B. (1983) *Imagined Communities: Reflections on the Origin and Spread of Nationalism*. Verso, London.

Bird, J. (1989) *The Changing Worlds of Geography*. Oxford University Press, Oxford.

Cloke, P., Philo, C. and Sadler, D. (1991) *Approaching Human Geography*. Paul Chapman, London.

Cooke, R.U. (1992). Common ground, shared inheritance: research imperatives for environmental geography. *Transactions, Institute of British Geographers*, **17**, 131–151.

Dear, M. (1988) The postmodern challenge: reconstructing human geography. *Transactions, Institute of British Geographers*, **13**, 262–274.

Deutsche, R. (1991) Boys town. *Environment and Planning D: Society and Space*, **9**, 5–30.

Development Education Centre (1992) *Developing Geography: a Development Education Approach at Key Stage 3*. Development Education Centre, Birmingham.

Domosh, M. (1991) Towards a feminist historiography of geography. *Transactions, Institute of British Geographers*, **16**, 95–104.

Enloe, C. (1990) *Bananas, Beaches and Bases: Making Feminist Sense of International Politics*. University of California Press, Berkeley.

Eyles, J. and Lee, R. (1982) Human geography in explanation.

Transactions, Institute of British Geographers, **7**, 117–122.

Giroux, H.A. and McLaren, P. (eds) (1994) *Between Borders: Pedagogy and the Politics of Cultural Studies*. Routledge, London.

Goodman, D. and Redclift, M. (1991) *Refashioning Nature: Food, Ecology and Culture*. Routledge, London.

Gregory, D. (1994) *Geographical imaginations*. Basil Blackwell, Oxford.

Hall, S. (1992) New ethnicities. In J. Donald and A. Rattansi (eds) *'Race', Culture and Difference*. Sage, London, pp. 252–259.

Harvey, D. (1989) *The Condition of Postmodernity*. Basil Blackwell, Oxford.

Harvey, D. (1990) Between time and space: reflections on the geographical imagination. *Annals, Association of American Geographers*, **80**, 418–434.

Jackson, P. (1989) *Maps of Meaning*. Unwin Hyman, London, (reprinted by Routledge, 1994).

Jackson, P. (1993) Changing ourselves: a geography of position. In R.J. Johnston (ed.) *The Challenge for Geography*. Basil Blackwell, Oxford, pp. 198–214.

Jameson, F. (1984) Postmodernism, or the cultural logic of late capitalism. *New Left Review*, **146**, 53–92.

Johnston, R.J. (1979) *Geography and Geographers*. Edward Arnold, London.

Johnston, R.J. (1983) *Philosophy and Human Geography*. Edward Arnold, London.

Johnston, R.J., Gregory, D. and Smith, D.M. (eds) (1994) *The Dictionary of Human Geography* (3rd edn). Basil Blackwell, Oxford.

Kincaid, J. (1988) *A Small Place*. Virago, London.

Larkin, N. and Widdowson, J. (1988) *Sweet or Sour?* Association for Curriculum Development, London (PO Box 563, London N16 8SD).

Livingstone, D.N. (1992) *The Geographical Tradition*. Basil Blackwell, Oxford.

Massey, D. (1991a) Flexible sexism. *Environment and Planning D: Society and Space*, **9**, 31–57; reprinted in D. Massey (1994) *Space, Place and Gender*. Polity Press, Cambridge, pp. 212–248.

Massey, D. (1991b) A global sense of place. *Marxism Today*, June, pp. 24–29; reprinted in D. Massey (1994) *Space, Place and Gender*. Polity Press, Cambridge, pp. 146–156.

Massey, D. (1994) *Space, Place and Gender*. Polity Press, Cambridge.

Naish, M. and Rawling, E. (1990) Geography 16–19: some implications for higher education. *Journal of Geography in Higher Education*, **14**, 55–75.

Naish, M., Rawling, E. and Hart, C.R. (1987) *Geography 16–19: the Contribution of a Curriculum Project to 16–19 Education*. Longman, Harlow.

Pile, S. (1994) Masculinism, the use of dualistic epistemologies and third spaces. *Antipode*, **26**, 255–277.

Radcliffe, S. (1994) (Representing) post-colonial women: authority, difference and feminisms. *Area*, **26**, 25–32.

Rose, G. (1993) *Feminism and Geography: the Limits of Geographical Knowledge*. Polity Press, Cambridge.

Sayer, A. (1989) The 'new' regional geography and problems of narrative. *Environment and Planning D: Society and Space*, **7**, 253–276.

Shields, R. (1991) *Places on the Margin: Alternative Geographies of Modernity*. Routledge, London.

Smith, N. (1993) Homeless/global: scaling places. In J. Bird, B. Curtis, T. Putnam, G. Robertson and L. Tickner (eds), *Mapping the Futures: Local Cultures, Global Change*. Routledge, London and New York, pp. 88–119.

Soja, E.W. (1989) *Postmodern Geographies: the Reassertion of Space in Critical Social Theory*. Verso, London.

Stoddart, D.R. (1986) *On Geography and its History*. Basil Blackwell, Oxford.

Stoddart, D.R. (1987) To claim the high ground: geography for the end of the century. *Transactions, Institute of British Geographers*, **12**, 327–336.

Stoddart, D.R. (1991) Do we need a feminist historiography of geography—and if we do, what should it be? *Transactions, Institute of British Geographers*, **16**, 484–487.

Unwin, T. (1992) *The Place of Geography*. Longman, Harlow.

Valentine, G. (1993). (Hetero)sexing space: lesbian perceptions and experiences of everyday spaces. *Environment and Planning D: Society and Space*, **11**, 395–413.

Developments in physical geography

7

Rita Gardner

Recent trends: recent reviews

The current state and the future of physical geography have
been the subject of a number of reviews since the 1970s,
commencing with that by Chorley (1971) in *Progress in
Geography* on 'The role and relations of physical geography'.
Most notable among reviews in the later 1970s and 1980s are
Brown's (1975) 'The content and relationships of physical
geography'; Clayton's (1985) 'The state of geography' and
Stoddart's (1987) 'To claim the high ground: geography for
the end of the century'. The most persistent theme through
these and the reviews in the 1990s (Clayton, 1991; Johnston,
1991; Cooke, 1992; Gregory, 1992; Newson, 1992; Goudie,
1994) is the fragmentation of physical geography through
the increasing development of specialisms. This has been
summarised as the pursuit of depth at the expense of
breadth. One response that comes through in several of the
reviews has been a call to 'speak out across subject boundaries'
(Stoddart, 1987, p. 334) and to concern ourselves once again
with an integrated approach to the environmental problems
of the world. This rallying call is as relevant to the
subdisciplines within physical geography as it is to physical
geography itself.

Physical geography is not unusual in being a subject
without clearly defined boundaries; but it is perhaps less

Geography into the Twenty-first Century, Edited by E.M. Rawling and R.A. Daugherty,
© 1996 John Wiley & Sons Ltd.

usual in the way it has evolved into a subject with a very diverse set of practitioners. Thus, in the first instance, a distinction can be made between developments in physical geography *per se* and developments in associated disciplines that assist in the understanding of physical features, environmental processes and environmental changes. It could be argued that many of the most interesting developments since the mid-1960s relevant to modern, global, physical geography come from attendant disciplines and are then applied to enhance geographical understanding. The second distinction, which is closely linked to the first, is between developments by those who would recognise themselves as physical geographers, by those who are physical geographers but call themselves by another name, and by those who genuinely contribute to physical geography from bases in other disciplines, such as sedimentology and engineering. This chapter takes the broadest view and it includes developments in associated disciplines that help to inform our understanding of biophysical environments; earth surface features; natural environmental resources; and environmental processes, degradation and change, both naturally and through human agencies.

In the more recent reviews several themes emerge. In a comparison of how the physical environment is changing compared with how physical geography is changing, Gregory (1992) concludes that contemporary physical geography is more in tune with recent changes in the environment than it was 20 years previously, most particularly with its increasing emphasis on global change and strategic (applied) research. However, he identifies a need for closer linkages between the study of dynamic change in the physical environment, of human-induced change and of their impacts together on global changes. Goudie (1994) broadly concurs with this view. He concludes his review by seeing hope for physical geography in that there has been a move away from compartmentalising phenomena and the subdivision of physical geography towards exploring interlinkages between humans and their environment, and applied problems.

Newson (1992), in contrast, argues that there is still a strong need to reclaim a broader base to physical geography; one which is prepared to integrate more with the social science aspects of geography, and which encompasses a holistic, conservationist perspective. He sees this as essential to the geographer's role in applied environmental management, but perceives, sadly, the continuation of an inevitable split between such 'environmentalists' and the pure school of 'reductionist' physical geography. He further argues that 'physical geographers must attract much of the blame for abandoning human geography, and the traditional broad aims of physical geography (e.g. landscape) in a rush to be imitators of geologists or engineers (or hydrologists) and to join their professional institutions' (Newson, 1992, p. 212). Cooke (1992) echoes similar themes when looking to the future: 'there is a danger that once again we shall allow our fissiparous tendencies to deflect our attention away from our difficult, but fundamentally important role on the common ground where human and physical geography overlap' (p. 132) and 'despite the distinctive importance of landscape in geographical training, it seems that geographers are retreating from the landscape and its inhabitants as a primary source of information' (p. 139). Others who have advocated the need to reverse the tendency to fragmentation in physical and environmental geography include Douglas (1987), Kates (1987) and Stoddart (1987). Thus, there clearly remains a difference of opinion among the 'great and the good' as to whether physical geography has commenced a reintegration around human–environment issues, or whether it continues along its recent historical pathway of fragmentation. The questions that follow are quite simple: Is there not room and, more importantly, respect enough for both approaches? Does not applied research need a sound base in pure science? What is to stop physical geography building on both its depth (pure) and breadth (applied) in the twenty-first century, providing the physical geographers among us are prepared to stand up and be counted as such?

In short, the dilemmas and worries that are clearly drawn out in these reviews concern breadth versus depth; holistic versus reductionist science; applied versus pure research; and integration versus fragmentation. Nearly all would seem to argue that there should be greater integration around the human–environment theme; the reviews differ only in the extent to which they recognise that this is now happening in terms of research. Thus, not only have the dilemmas remained the same since the 1970s, and some might argue that they are characteristic of the postmodern era, but today they are reflected even more widely in the future funding of research. The traditions of reductionist pure science formerly embraced with fervour by the Research

Councils are being set against the promotion of integrating strategic research focused around wealth creation, solving environmental problems and improving the quality of life by both the UK government and the EU. How ironic that this long-lasting debate of the physical geographers should have become a national debate as to the future role of science and scientific research.

Fragmentation: funding and the role of technology

That there has been since the mid-1970s increasing differentiation and fragmentation of physical geography is not disputed. This has come about through a number of largely external forces that have driven the research agendas and perceptions of practising physical geographers, rather than through a consensus within the 'subject' and its practitioners as to future form and direction. Repeated calls from within for reunification of the subject around a core, as summarised above, appear to have had little impact. Among the stronger external forces have been the approaches to 'pure' reductionist research adopted by the Research Councils, which were in turn informed to some extent by wider social and political agendas, and by power groups that developed within those Councils. Physical geographers have long suffered from two drawbacks with respect to the Research Councils. The first is that their expertise is spread across several different Research Councils and even across several committees within the Councils, thereby preventing them from forming a powerful lobby. Secondly, the discipline has been perceived by others, and practitioners have, to some extent, come to believe themselves, that physical geography is not a first-rate earth science. These forces strengthened trends towards a close alliance with 'respectable' allied disciplines, into adopting the labels of those disciplines, and into inevitable fragmentation of the 'subject'. There were undoubtedly many individual benefits in terms of funding, research expertise and contacts to be gained by doing so, but it has done little to strengthen the external image of physical geography, and there remains a substantial image-building exercise to be done in this area. There is a glimmer of hope that this may be under way following the recent success of physical geographers.

Research Council funding was also partly responsible for the increasing dominance of geomorphology within physical geography, and for the development of the hydrological aspects of geomorphology above all others in the UK. Emphasis on Quaternary studies of environmental change has also increased in importance recently, fuelled partly by the recognition that past environmental changes can provide analogues for possible future changes, and by an increasing awareness of climate change. However, research by physical geographers into climatology and meteorology, periglacial geomorphology, pedology, and coastal processes is now very limited in the UK.

Clearly, the trend towards fragmentation is not simply a function of the Research Councils. There were other forces at work too, including the information explosion, the nature of undergraduate training, the wider scientific norms of the time, the need for research image seen in the proliferation of 'centres' and 'units', and technological developments. Whatever the cause, physical geography has increasingly come to resemble a polo mint. The debate as to whether this is perceived as good or bad for the 'subject' is largely academic and arguably not productive. It exists, and the profession needs to go forward from here.

Within the fragmented subject, research developments in the 1980s and the 1990s have probably been influenced more by technological change and advance than by any other single factor. In most cases the technological developments have been imported from other disciplines, especially physics, chemistry, engineering and computer science, and applied within a geographical setting. While the new technologies have served to expand the physical geographers' toolkits, in common with those of other earth scientists, often the geographer does not have the necessary science or mathematical skills to develop further the technologies, nor the time in the present research environment to acquire the skills so to do. One of the clearest examples is in global climate circulation models. We make very effective use of the model outputs (e.g. Parry, 1990) but have relatively little input into the development or refinement of the models themselves.

Of the many technological developments that could be cited, there are perhaps six that stand out above the others as influencing the course of research in physical geography. The first is the improved capacity and increased flexibility in continuous and regular monitoring of environmental processes and fluxes. The widespread dissemination of automated logging systems, and the ability to transmit logged information to end users directly by satellite, have greatly expanded the efficiency and speed of collection of field data. Their accuracy improves and the volume of data increases most particularly from the less developed locations, provided the technology can be sensibly maintained. Not only has such information improved understanding of processes themselves, their rates of operation, and their spatial and temporal variability, but it has assisted in the derivation of empirical models (e.g. Morgan et al., 1984) and in providing field information for the parameterisation and validation of process-based models. The future for this type of research is becoming less certain, partly because of the time and costs involved, partly because empirical research has become less fashionable and, sadly because of the continuing difficulty of obtaining funding for long-term monitoring studies.

The second technological improvement is in data storage and manipulation facilities offered by digital computers, especially with the increasing capacity of PCs, and the flexibility of off-the-shelf software. In particular, this has opened up wider possibilities for computer modelling of earth surface processes, and it has extended them to a larger range of researchers. The more complex, physically based models involve the solution of partial differential and continuity equations that describe the processes involved and that incorporate the laws of conservation of mass and energy (e.g. Lane et al., 1992). Examples include the development of distributed models for slopes and catchments, in which inputs and outputs of matter are determined across individual cells in both vertical and horizontal dimensions. It is also possible to integrate submodels that deal with specific components of the system. Such modelling, it is argued, improves prediction of behavioural patterns for practical applications, allows detailed examination of process interactions, and enables scenario-setting through computer simulation of changing boundary conditions (Anderson, 1988). Physical geographers have generally been more concerned with the use of models to examine process interactions (Parsons and Abrahams, 1992). However, some concern has been expressed that their complexity and apparent sophistication can mask fundamental weaknesses in the algorithms on which they are based (e.g. Grayson and

Moore, 1992) and that such modelling has a long way to go before it develops its full potential. At present the models are often attuned to short time and small spatial scales and one of the challenges for the future is to upscale the models to larger grid scales of hundreds of kilometres. This shift from empirical measurement to modelling is not simply a function of the changing technology; the technology has facilitated the shift within a receptive research environment where modelling is currently seen as fashionable and where it offers the advantages of relatively low costs, and quick returns. As Cooke (1992) suggested, we 'seem increasingly to be trapped by our VDUs and their fleeting images' (p. 139).

Global environmental monitoring has become a reality thanks to developments in remotely sensed satellite monitoring. It is now able to provide synoptic views of environmental, geomorphological and vegetational phenomena, and it has taken the physical geographer from the catchment scale into the global realm, albeit leaving the researcher with difficulties of linking between the different scales. Examples of remote sensing applications in physical geography range from land resource inventories and the mapping of land cover change (including deforestation) using Landsat MSS, and polar ice cover change using Seasat (Drewry et al., 1991); to monitoring of sulphur dioxide clouds from volcanic eruptions (Bluth et al., 1992) using TOMS (total ozone mapping spectrometer); and oil pollution mapping (Cross, 1992) using, for example, AVHRR (advanced very high resolution radiometry) images. When combined with global models, Gregory (1992) argues that remote sensing has catalysed progress at the global scale. Earth observation and monitoring are now clearly on the agenda for the future with its recognition in the new structure of the Natural Environment Research Council (NERC).

Geographical information systems have vastly improved the capacity for the storage, manipulation, presentation and analysis of spatially referenced data; and have further offered the physical geographer the possibilities of using digital elevation models for data representation and modelling. Until recently, much of the use of this flexible toolkit has been in planning and mapping of utilities and services. Environmental applications, other than simply mapping the distribution of physical characteristics such as soils, land use, etc. are still in their infancy. Recent applications to terrain modelling (e.g. Stocks and Heywood, 1994), soil erosion hazard mapping (Griffiths and Richards, 1989) and modelling (De Roo et al., 1989), and to modelling landslide distribution (Wang and Unwin, 1992) are clear indicators of the potential future role of GIS at local and regional scales of analysis.

Since the mid-1970s, advances in sediment dating techniques, and in the interpretation of stable (carbon and oxygen) isotope ratios, have brought very substantial developments in understanding the frequency, magnitude, spatial variability and timing of natural climate change and, have provided a global basis for subdividing the Quaternary. Some isotopes, such as lead-210, have also added greatly to our understanding of the timing and rates of environmental change resulting from recent human activities. This whole research area continues to be very productive into the 1990s. For example, the recent ability to date a wider range of terrestrial and coastal silicate sediments using developments in the luminescence techniques, has opened up the mid to late Quaternary deposits of the continental interiors to reappraisal both in terms of timing and process of their formation, and in the information they provide on regional climate change (e.g. Rendell et al., 1993). Uranium series dating of Fe/Mn oxyhydroxide accumulations in soils is useful in dating terrestrial deposits and palaeosols in the tropics. The revival of interest in Quaternary environmental change, in general, has also stemmed from our concerns with sustaining the future environment and the need to understand past changes both because they may well be repeated in the future, and because they provide the best analogies, at 9000 years BP, to the likely future effect of global warming (Roberts, 1989). Some of the most outstanding work of late has shown the rapidity with which climate change can occur naturally. For example, records of the last interglacial in ice cores suggest that changes of the order of 5–10 °C can occur within a few tens of years and persist for hundreds and thousands of years (Greenland Ice-Core Project (GRIP) members, 1993). This development in knowledge has been facilitated by high resolution dating techniques, and the rate puts natural climate change on a par with the possible effects of human-induced global warming, both in terms of rapidity as well as magnitude.

There is a wide range of other specialised analytical techniques that have opened doorways to new interpretations of old problems and to the analysis of new problems arising from human interactions with the environment. The use of diatoms as indicators of water quality has found wide applicability, most notably in the studies of surface water acidification resulting from 'acid rain' (Battarbee and Charles, 1988). Another example is the isotope caesium-137, which is offering opportunities to examine soil erosion and soil redistribution on hillslopes since the mid-1960s (Walling and Quine, 1991), a time when increasing human pressure resulting from population expansion is being felt in many of the most sensitive environments. This may be complemented by the use of mineral magnetic properties of accumulated sediments to assess the sources of eroded material, both for recently eroded sediments and in older lacustrine materials (Oldfield and Clark, 1990). All of these methods contribute to understanding change in the recent historical past—the last 150 years—the importance of which was most lucidly argued by Cooke (1992).

Many of these techniques have clearly offered opportunities for physical geographers to contribute more fully to applied research relating to some of the major environmental and resource issues of our time—desertification, global warming, water acidification, air pollution, deforestation, soil erosion and fertility management, flooding, and wetland conservation. As Goudie (1994) and Gardner and Hay (1992) note, there has been a welcome increase in the contribution to applied research in recent years, including the field of environmental implications of development. Most of this has been in terms of improved understanding of the impacts of human activity on the environment, and the investigation of future scenarios (e.g. Parry, 1990). However, we are yet to see much direct translation of physical geography research findings into environmental management policy. Arguably, this translation may be for others to do, nevertheless, the discrete boundary between environmental manager/policy-maker and academic researcher has blurred little since the mid-1970s. There are, of course, some notable exceptions including that of nitrate pollution management (Burt et al., 1993).

We can only speculate as to the reasons for this relative failure in the policy arena. Is the research not sufficiently geared to policy priorities; is it because of our ineffectiveness, as academics, at developing links with policy makers and other user groups, and the perceived undesirability in terms of peer prestige of leaving the ivory tower; or the credibility of geography in the wider world, together with the apparent relative merits of commercial consultancy over academic research; or the unwillingness of academics to generalise from uncertainty, complexity and, frequently, limited data to 'on the ground' solutions; or the sheer responsibility; or the scientific tradition of imposing knowledge rather than developing solutions with and within the institutional frameworks that exist in the problem locations. Whatever the cause, it highlights a difference between the role and relevance of geography as sometimes taught at secondary level, and the contribution that geographical research presently makes to solving pressing environmental problems.

It has been argued that geographers have unique opportunities in environmental management at a time of dramatically increased, although possibly now waning, environmental awareness since the mid-1970s. The geographers' understanding of environmental processes together with their awareness of social, political and economic forces is generally alluded to in this context. But real links between human and physical geography are as tenuous now as they have ever been and their research cultures are widely divergent. The renewed interest in finding multidisciplinary research may improve the possibilities for collaboration in future. However this may have come too late; one is left pondering whether physical geographers have been left standing at the bus stop as the ecologists, hydrologists, earth scientists and 'environmental scientists' have rushed to climb on board with their polished environmental images and newly established environmental management institutions (e.g. IWEM and IEEM). Sadly, there is little evidence since the mid-1970s of physical geographers grasping the environmental management initiative with sustained enthusiasm and commitment.

The conceptual frameworks: multiple and consistent

The conceptual frameworks within which physical geography operates have not undergone revolutionary change since the 1960s and 1970s, although advances in knowledge within these frameworks may have been substantial. Plate tectonics—arguably the closest earth sciences has come to revolution since the turn of the century—continues to provide the unifying framework for the evolution of earth surface features at the global and continental spatial scales and over geological time-scales. Within this context, the potential importance of tectonics, and in particular of the Himalayan orogeny, in bringing about global cooling and signalling the onset of the Quaternary ice age, is presently sending ripples through the earth science community (Raymo and Ruddiman, 1992), but there remains much still to be done to substantiate the hypothesis (Berner, 1992). The years since the mid-1970s have also seen greater recognition of the importance of Quaternary and older tectonic activity in affecting erosion, sedimentation and long-term landform evolution (Summerfield, 1991).

While considerably more activity, in terms of research output since the mid-1970s, has occurred in the area of Quaternary environmental change, and particularly change over the past 100 000 years, the framework for such studies goes back to stratigraphic principles established by Lyell in the nineteenth century. However, the research emphases within this framework have shifted. For example, there has been a greater focus on establishing 'absolute' time-scales for the regional stratigraphies, the early need for which was argued most convincingly by Vita-Finzi (1973), and in developing a global stratigraphic subdivision of the Quaternary period based on the isotopic and biological characteristics of deep-sea core sediments (as discussed above). Correlations between the regional and global stratigraphies are increasingly possible, and the chronostratigraphic studies have, in turn, provided substantial support for the 'Milankovitch hypothesis' as a prime (Broecker and Denton, 1990), but not undisputed (Broecker, 1992) cause of cyclic climatic change within the Quaternary.

Similarly, in the third main dimension of physical geography, that of current environmental dynamics, the systems approach still dominates as a unifying framework for research and teaching. Although, as pointed out by Newson (1992), physical geographers have concerned themselves almost exclusively in research with the study of fluxes and processes, and have effectively turned their backs on the wider possibilities offered by systems theory. Within this narrow interpretation, there has developed a much clearer conceptual and practical understanding of thresholds, complex response, relaxation, efficiency, and sensitivity to change, and the energetics of certain systems are better defined (Gregory, 1987), although sensitivity to human impact is often not well understood (Goudie, 1994). Moreover, there has been increased interest by geomorphologists in the nature of sediments and in the interactions between fluids and materials, thereby helping to reduce even further the artificial boundary between geomorphology and sedimentology—a fact that is reinforced in the joint affiliation of the British Geomorphological Research Group (BGRG) with both the RGS/IBG and the Geological Society. In the field of glacial geomorphology, for example, sediments are being used to reconstruct glacial dynamics, including glaciotectonic processes (Croot, 1988). Nevertheless the scientific basis underpinning physical geography has not developed as much as was hoped (Gregory, 1992), and the links between understanding processes and long-term landform evolution still often remain tenuous. Thus, while the spatial scale and intellectual depth of hydrological, geomorphological and some pedological process studies may have changed, the narrowly defined interpretation of the systems framework remains at the forefront of the area.

Lastly, in a fourth dimension, biogeography continues to derive its framework from ecology and ecological principles. It is currently seeing something of a revival as issues of biodiversity and conservation move high up the political agenda. For some, physical geography has been excessively dominated by geomorphology (Brown, 1975; Simmons, 1978), and the research and funding imbalance between geomorphology and other aspects of physical geography is a trend which still continues (Gardner and Hay, 1992). The development of biogeomorphology—wherein the importance of organisms and soils in affecting geomorphological processes is finally receiving recognition (Viles, 1988)—is helping to bridge this divide. But some would go so far as to see biogeography as a preferable core to physical geography (Stoddart, 1986). The wider extension of ecological principles to other areas of geography has also been advocated. As early as 1971, cultural ecology was suggested as a conceptual framework for human/environment studies (Cooke, 1971), and more recently ecology is seen as offering a promising framework for applied environmental geography. Certainly, there is at present no clear intellectual framework for studies of human/environment interactions, at a time when there is considerable uncertainty and lack of knowledge over the likely outcomes of such interactions on both the climate and land resource (Ives and Messerli, 1989).

Decoupling of approaches to secondary and tertiary teaching in physical geography?

Reflecting in part the specialisation in research, tertiary-level teaching has become entrenched, for the most part, into systematic subdisciplines. Nevertheless, courses integrated around human/environment themes and specific environments of the world still exist, or have been recently reintroduced, in some institutions. Using systematic knowledge such courses have come a long way from the traditional 'regional' geographies of the 1950s and 1960s. In contrast, geography at secondary level has taken an increasingly holistic approach, involving integration across geographical subdisciplines and often focused on problem-solving and decision-making (see Davidson and Mottershead, Chapter 18). Ironically, the secondary approach is usually hampered, inevitably at this stage, by poor detailed knowledge of underlying processes; whereas the entrenched subdisciplines at the tertiary level fail to capitalise on the greater, but compartmentalised, knowledge of the students and the opportunities that this offers for integration. As a result, an integrated and broadly based geographical education at university—one that encompasses effectively both breadth and depth—is unusual in the current climate of higher education, although the often large number of options of a modularised programme is not without its reward. This situation has resulted from several well-known contributing causes, including constraints imposed on university teachers by other demands on their time, high staff–student ratios, the need for course flexibility, the appeal and marketability of wide choice to prospective students, administrative ease, ideas of 'progression', educational fashion, and the (highly sensible) recognition that we should teach to our research specialisms. In so teaching at tertiary level we are, however, often perpetuating fragmentation of the subject.

From the university perspective, the dichotomy of approach between teaching at secondary and tetiary levels seems to be widening. This decoupling of school and university physical geography can result in the students being less well prepared for current undergraduate courses and for the approach to the subject taken in many such courses. A commonly heard complaint is that A level students know so much less than they ever used to; but perhaps as university teachers we do not always recognise sufficiently that they come to us with different skills and expectations.

Conclusion

The 1960s saw the emergence of two new, exciting conceptual frameworks relevant to physical geography, namely the general systems theory and plate tectonics. Both of these have stood the test of time, although physical geography has tended to exploit only some aspects of each and, most notably, to have largely turned its back on control systems. The years since 1970 have been witness to major consolidation and advances in knowledge within these newly established conceptual frameworks, and within the longer-established frameworks borrowed from ecology and stratigraphy. Our advance in depth of understanding has been monumental, and it has been facilitated by, and in turn further encouraged, the fragmentation of physical geography and of its subdisciplines, especially geomorphology. But developments have been strongly biased—towards geomorphology and Quaternary environmental change—and away from soils, climatology and meteorology which have been relinquished largely to the realms of soil science, mathematics and physics. Meanwhile, biogeography has kept a firm foot in the physical geography camp, and is now seeing a revival as issues of biodiversity have come on to the political agenda. All these shifts are perhaps best interpreted as physical geography identifying and consolidating its core in a more scientific era.

Many of the developments within the fragmented areas have been facilitated by the application of newly emerged technologies for continuous monitoring, dating, computer mapping, remote sensing, modelling and data storage/manipulation, and the identification of new forms of environmental proxy data. Many of the fragmented subdisciplines are closely aligned with bodies outside physical geography, and the fragmentation has been further encouraged by approaches to research funding. As the discipline has become fragmented, ironically so the scale of research has expanded to regional and global proportions, owing to opportunities offered by satellites.

Views as to the desirability of fragmentation vary, although there is a broad consensus among the recent reviews of the need for physical geography to reassert its integrating core around a human/environment theme, at a time when environmental issues worldwide and in the UK are still relatively high on the political agenda. There is little to show that such calls have stimulated a positive response in the past, and the recent increase in applied environmental geography research probably owes as much, if not more, to funding opportunities than to rallying from the elder statesmen of the subject. Research of this type is often hampered by shortage of good data, and the fact that such data are expensive to collect and may require long-term monitoring programmes. Moreover, human/environmental research is still searching for a conceptual framework.

With the new emphasis on *multidisciplinary* applied research, together with increasingly technological sophistication, it is arguable whether the geographer will ever be able to dominate effectively the 'central ground' of human/environment interactions, especially as so many other disciplines have seen the environmental light and are already well established on the 'bandwagon'. But without doubt physical geographers need to continue to assert their human/environmental credentials, to cross the boundaries between fragmented subdisciplines, and to have greater input into management and policy. Above all we need to allow ourselves a flexible future that builds on our strengths: a future where we continue to develop our research depth and diversity, but where we do not lose sight of, and can also reassert, our breadth in the human/ environment area.

References

Anderson, M.G. (ed.) (1988) *Modelling Geomorphological Systems.* Wiley, Chichester.

Battarbee, R.W. and Charles, D.F. (1988) The use of diatom assemblages in lake sediments as a means of assessing the timing, trends and causes of lake acidification. *Progress in Physical Geography,* **12**, 552–580.

Berner, R.A. (1992) Palaeo-CO_2 and climate. *Nature,* **358**, 114.

Bluth, G.J.S., Doiron, S.D., Schnetzler, C.C., Kruegar, A.J. and Walter, L.S. (1992) Global tracking of the SO_2 clouds from the

June 1991 Mount Pinatubo eruption. *Geophysical Research Letters,* **19,** 151–154.

Broecker, W.S. (1992) Upset for Milankovitch theory. *Nature,* **359,** 779–780.

Broecker, W.S. and Denton, G.H. (1990) What drives glacial cycles? *Scientific American,* **169,** 43–50.

Brown, E.H. (1975) The content and relationships of physical geography. *Geographical Journal,* **141,** 35–48.

Burt, T.P., Heathwaite, L. and Trudgill, S.T. (eds) (1993) *Nitrates: Processes, Patterns and Control.* Wiley, Chichester.

Chorley, R.J. (1971) The role and relations of physical geography. *Progress in Geography,* **3,** 89–109.

Chorley, R.J. and Kennedy, B. (1971) *Physical Geography: a Systems Approach.* Prentice-Hall, London.

Clayton, K.M. (1985) The state of geography. *Transactions of the Institute of British Geographers,* NS **10,** 5–16.

Clayton, K.M. (1991) Scaling environmental problems. *Geography,* **76,** 2–15.

Cooke, R.U. (1971) Systems and physical geography. *Area,* 3, 212–216.

Cooke, R.U. (1992) Common ground, shared inheritance: research imperatives for environmental geography. *Transactions of the Institute of British Geographers,* NS **17,** 131–151.

Croot, D.G. (ed.) (1988) *Glaciotectonics.* Balkema, Rotterdam.

Cross, A.M. (1992) Monitoring marine oil pollution using AVHRR data: observations off the coast of Kuwait and Saudi Arabia during January 1991. *International Journal of Remote Sensing,* **13,** 781–788.

De Roo, A.P.J., Hazelhoff, L. and Burrough, P.A. (1989) Soil erosion modelling using 'ANSWERS' and Geographical Information systems. *Earth Surface Processes and Landforms,* **14,** 517–532.

Douglas, I. (1987) The influence of human geography on physical geography. *Progress in Human Geography,* **11,** 517–540.

Drewry, D.J., Turner, J. and Rees, W.G. (1991) The contribution of Seasat to ice sheet glaciology. *International Journal of Remote Sensing,* **12,** 1753–1754.

Gardner, R. and Hay, A.M. (1992) Geography in the United Kingdom, 1988–1992. *The Geographical Journal,* **158,** 13–30.

Goudie, A.S. (1994) The nature of physical geography: a view from the drylands. *Geography,* **79,** 194–209.

Grayson, R.B. and Moore, I.D. (1992) Effect of land surface configuration on catchment hydrology. In A.J. Parsons and A.D. Abrahams (eds), *Overland Flow: Hydraulics and Erosion Mechanics,* UCL, London, pp. 147–176.

Gregory, K.J. (ed.) (1987) *Energetics of Physical Environment: Energetic Approaches to Physical Geography.* Wiley, Chichester.

Gregory, K.J. (1992) Changing physical environment and changing physical geography. *Geography,* **77,** 323–335.

Griffiths, J.S. and Richards, K.S. (1989) Application of a low-cost database to soil erosion and soil conservation studies in the Awash Basin, Ethiopia. *Land Degradation and Rehabilitation*, **1**, 241–262.

GRIP (Greenland Ice-Core Project) members (1993) Climate instability during the last interglacial period recorded in the GRIP ice core. *Nature*, **364**, 203–207.

Ives, J.D. and Messerli, B. (1989) *The Himalayan Dilemma*. Routledge, London.

Johnston, R.J. (1991) A place for everything and everything in its place. *Transactions of the Institute of British Geographers*, NS **16**, 131–147.

Kates, R. (1987) The human environment: the road not taken, the road still beckoning. *Annals of the Association of American Geographers*, **77**, 525–534.

Lane, L.J., Nearing, M.A., Laflen, J.M., Foster, G.R. and Nichols, M.H. (1992) Description of the US Department of Agriculture water erosion prediction project (WEPP) model. In A.J. Parsons and A.D. Abrahams (eds), *Overland Flow: Hydraulics and Erosion Mechanics*, UCL, London, pp. 377–392.

Morgan, R.P.C., Morgan, D.D.V. and Finney, H.J. (1984) A predictive model for the assessment of soil erosion risk. *Journal of Agricultural Engineering Research*, **30**, 245–253.

Newson, M. (1992) Twenty years of systematic physical geography: issues for a 'New Environmental Age'. *Progress in Physical Geography*, **16**, 209–221.

Oldfield, F. and Clark, R.L. (1990) Lake sediment-based studies of soil erosion. In J. Boardman, I.D.L. Foster and J.A. Dearing (eds), *Soil Erosion on Agricultural Land*, pp. 201–228.

Parry, M.L. (1990) *Climate Change and World Agriculture*. Earthscan, London.

Parsons, A.J. and Abrahams, A.D. (eds) (1992) *Overland Flow: Hydraulics and Erosion Mechanics*. UCL, London.

Raymo, M.E. and Ruddiman, W.F. (1992) Tectonic forcing of late Cenozoic climate. *Nature*, **359**, 117–122.

Rendell, H.M., Yair, A. and Tsoar, H. (1993) Thermoluminescence dating of periods of sand movement and linear dune formation in the northern Negev, Israel. In K. Pye (ed.), *The Dynamics and Environmental Context of Aeolian Sedimentary Systems*, Geological Society Special Publication 72, pp. 69–74.

Roberts, N. (1989) *The Holocene: an Environmental History*. Basil Blackwell, Oxford.

Simmons, I.G. (1978) Physical geography in environmental science. *Geography*, **63**, 314–323.

Stoddart, D.R. (1986) *On Geography*. Basil Blackwell, Oxford.

Stoddart, D.R. (1987) To claim the high ground: geography for the end of the century. *Transactions of the Institute of British Geographers*, NS **12**, 226–327.

Stocks, A.M. and Heywood, D.I. (1994) Terrain modelling for

mountains. In M.F. Price and D.I. Heywood (eds), *Mountain Environments and Geographic Information Systems*, pp. 25–40.

Summerfield, M. (1991) *Global Geomorphology*. John Wiley, Chichester.

Viles, H.A. (ed.) (1988) *Biogeomorphology*. Basil Blackwell, Oxford.

Vita-Finzi, C. (1973) *Recent Earth History*. Academic Press, London.

Walling, D.E. and Quine, T.A. (1991) The use of caesium-137 measurements to investigate soil erosion on arable fields in the UK: potential applications and limitations. *Journal of Soil Science*, **42**, 147–165.

Wang Shu-Quaing and Unwin, D.J. (1992) Modelling landslide distribution on loess soils in China: an investigation. *International Journal of Geographical Information Systems*, **6**, 491–505.

Environmentalism and geography: a union still to be consummated

8

Tim O'Riordan

Introduction

In the space of 25 years environmental issues have entered the day-to-day lives of everyone. Admittedly there is talk of green fatigue and a 'contrarian backlash' (Beckerman, 1995; North, 1995). The claim is that environmental concerns do not stand up to the heat of recession or to unemployment and economic insecurity generally. In the past, this was the case. The first great environmental era, namely the progressive conservation movement in the United States between 1890 and 1910 (Hays, 1957), ended in the economic disarray of the early 1920s. The second conservation movement, again in the US in the mid-1930s, also crashed in the war and its reconstruction aftermath. In both cases the drive for wealth creation, and the political demands for more real household income, pushed environmental concerns, such as they were, to one side. There was very little environmental geography in those days: the matter was mainly of interest to historians (e.g. Hays, 1957, 1987) and to ecological philosophers (Leopold, 1949).

Today it is very different, and do not be put off by the fashionable, but short-lived, anti-environmentalist rhetoric of the present day. Environmental matters are enshrined in many laws in all countries: regulation over environmental protection and health and safety is becoming more effective:

Geography into the Twenty-first Century, Edited by E.M. Rawling and R.A. Daugherty,
© 1996 John Wiley & Sons Ltd.

international treaties require nations to meet global objectives when they certainly would not do so on their own; business is expected to undertake audits of its environmental and social impacts and to demonstrate how it will reduce these adverse effects by publishing policies and targets; and national governments are beginning to do the same in their so-called sustainability audits. Even in education, environmental issues are having some recognition. (For a regular review see *ENDS*.[1]) Admittedly, the pace and degree of real progress are far from what is desirable if environmental and social well-being is a legitimate good. But the progress is unmistakable.

The tragedy is that the subject matter is not being promoted by geographers, who should have cornered the field years ago. Nor are many geography departments leading the way in contemporary environmental politics. This is especially absent in terms of both global research programmes and European issues. The EU is by far the most active environmental organisation of its kind anywhere, yet few geographers are taking an interest in what should be a central theme of all aspects of the geography of the continent (Sands, 1994; Cameron et al., 1995; Haigh and Lanigan, 1995).

One should ask why geographers have been so slow in developing the high profile that is needed in environmental research. Four reasons can be suggested:

1. Geographers have maintained the unfortunate separation of the physical and human aspects of the discipline, and if anything, have increased the intellectual gap that divides these manifestly integratable approaches.
2. The exciting margins of modern environmental research are at the edges of disciplines some distance removed from the heartland of geography; in chemistry, population dynamics, chaos theory, economics, philosophy, politics and the law. It is only with a good grounding in the core of these disciplines that the potential for innovation at their margins of commonality can be most appreciated.
3. The mixture of the analytical (using models, numbers and statistics), and the judgemental (using principles of politics, philosophy and sociology) should come easily to geographers, but does not. A special training is required to incorporate, sequentially and episodically, these two approaches into a unified strand.
4. The modern version of environmentalism, as we shall

see, captures both the social and political agendas, yet geographers, who should be best placed to snap up this opportunity, still remain focused on concepts such as gender, inequality, locational disadvantage, or carrying capacity for threatened habitats.

The contemporary environmental scene is very active, very innovative, very integrative (both in method and subject matter), and highly political. It should be the heartland of a modern geography course, connecting technique and enquiry into a lively discourse, and providing a major contribution to the task of preparing pupils for a world of considerable challenge. The time has certainly come for geographers to grasp this message. Otherwise, sadly, they will be left behind in a world that will look elsewhere for intellectual leadership and methodological guidance.

Modern environmentalism

The environmental idea is driven by the recognition of those persuasive but irreconcilable contradictions that have always been part of the human condition.

A desire to control nature for power and wealth; yet a need to protect nature for ultimate survival

This is the age-old dilemma between 'development' and 'preservation', between optimistic progressivism and precautionary stewardship, between radicalism and conservation, between individualism and egalitarianism that lies in all of us to very varying degrees. It is often tempting to divide societies into the 'bad guys and the good guys' in this respect. That would be extremely misleading. Both strands contain forces of good and bad. Both are vital for the advancement of the human condition. Both only reveal themselves in conflict and temporarily successful resolution or mediation. The environmental movement needs its crises to provide it with zones of debate and attention, and to create the incentive to reform unsatisfactory and ill-suited procedures. As the great philosopher-scientist Eric Ashby (1977, pp. 85–86) put it so beautifully:

. . . conflicts of values in environmental policy turn up
again and again . . . people are not indifferent to these
conflicts, nor do policymakers override them. The
conflicts themselves are in my view immensely useful,
for they provide a continuing debate about moral
choice: choice between hard and soft values, choice
between indulgence in the present and consideration
for the future. They oblige people to strike a balance
between counting what can be quantified and caring
for what cannot be quantified. Every choice redefines
the goal for environmental policy. In the process of
choice the protection of nature is becoming more
securely implanted into the culture. . . .

Bennett (1992) produced a valuable series of case studies to
illustrate the dilemmas of frustrating irreconcilability in
environmental disputes. His thoughtful essays are well
worth reading. They show how power is deployed head on
by sheer muscle, and, much more subtly, by persuading
people that their 'real' interests lie in destroying the natural
world and their weaker and politically disenfranchised
human brethren, in the name of 'progress' and 'security'. It
is no wonder we live in a non-sustainable age: we have been
marketed, believe it or not, actually to fight the transition to
survival for the human species as a whole. The key institutions
of economics, politics, the law and cultural relations all bias
the balance of argument in the direction of non-sustainability.

The characteristic of democracies to vote for short-term goods, rather than for avoidance of long-term loads

The environmental game is located in favour of the present
generation. We have no scientific, cost–benefit, legal or
democratic apparatus for properly taking into account
potential disasters that could cost future generations very
dear. There is, in short, no geography of the meso-scale or
what might happen at the end of our average lifetimes,
about 50 years from now, and beyond. This is also a paradox
of the so-called sustainability transition. Democracies vote
out governments that promise jam tomorrow but stale bread
today; yet without democratic consent the awesome task of
switching a non-sustainable society into a more sustainable
pathway is doomed to counter-productiveness. As we shall

see, we need a new vocabulary, fresh mediating techniques, and better integration of imagination into computer models, if we are going to make progress. This will be one of the many tasks of the 'geography of envisioning' (see Milbrath, 1989 for an early statement, and O'Riordan et al., 1993 for a more practical example).

The way in which managerialism promotes efficiency at the expense of fairness and redistribution

One of the great dilemmas of the modern age is the battle between competitive advantage based on efficiency of operation and the concern for those who lose in the process *because they are already losers*. Consider, for example, the current concern with environmental taxation (e.g. Turner et al., 1993). All resource exploitation is underpriced, partly because it is often deliberately subsidised to pay the already wealthy to be profligate (Brown and Pearce, 1994 show this with regard to tropical forest depletion), and partly because the environmental and social costs associated with extraction, transportation and ultimate disposal are not paid for by those who cause them. These so-called external, social or third-party costs tend to fall disproportionately upon the poor, or the politically disorganised (see e.g. Bullard, 1994 for evidence of this with regard to waste disposal). Now, if a tax is levied to incorporate these costs, then it is likely that the higher prices will be unfairly borne by the poor, who often have little choice but to pay up. For example, a carbon tax aimed at correcting the uncompensated cost of creating global warming (see e.g. Hayes and Smith, 1993; Frankhauser, 1995) would fall heavily on the poor who pay four times as much in relative income terms on carbon-based feed than do the rich (Pearson and Smith, 1991).

Economics, as a philosophy, has not been as good as it ought to be about incorporating equity, or distributional and compensatory measures, into its theories of efficiency and optimality. The theory assumes that, if those who do gain could pay off those who lose and still be better off, then a resource-use decision, on minerals extraction, forest depletion or water diversion should take place (Turner et al., 1993).

All this assumes that such costs can be identified, that all those who lose are also recognised, and that there is a mechanism of transfer payments. In fact, none of this is effectively in place, so the politics of modern environmentalism are characterised by two sets of gainers—*polluters* and those depleting natural resources in net terms; and *social disruptors*, those who remove jobs or deny people economic security so they are thrown on the mercy of the state or their family providers to survive. Both sets of gainers who operate in the name of 'least cost management' under the guise of efficiency, are actually creating a series of losers, many of whom are already suffering. Again, there is no adequate mechanism for identifying and overcoming this, nor can science always help. Because science is not capable of dealing with funda-mental unknowns, and because science cannot adequately integrate physical alteration into social values and ethical principles, rooted in fair democratic procedures, so the 'double non-dividend' becomes a tragic but familiar feature of modern environmentalism. It is no wonder that those who suffer so much are trying to organise themselves around fundamental principles of 'ecological civil rights' through which a guarantee of survival for both humans and ecosystems becomes enshrined in international law (Brown Weiss, 1989).

The tendency for property rights to protect private owners but to penalise the public interest in property

The law of property is very good on protecting the interest of the owner, who has rights to reasonable use, and to the avoidance of nuisance or 'taking' of the private enjoyment of land. But all property, whether it be a home or a landscape, functions to varying degrees in the public interest. This may be a result of appearance, the buffering effect it provides against physical damage, or the role of associated ecosystems in removing degradable substances or sequestering pollutants. For example, the ocean surface is a very lively zone for emitting and removing gases, many of which are of immense importance in controlling climate. Yet that surface can be damaged by pollution or possibly by excessive ultraviolet radiation caused by the removal of stratospheric ozone, or by usurpation by those who get in first. The US and former Soviet Union countries between them emitted nearly two-fifths of all the carbon dioxide (CO_2) that the world creates. In so doing, 7 per cent of the world's population have absorbed nearly half of the global carbon sinks. These sinks are not only for everyone to use. They are also the essential mechanism by which the planet copes with excess carbon. For such a disproportionate 'taking' of a collective human right, the perpetrators paid nothing for the privilege. Again there is no fair mechanism that ensures that such life-preserving 'common' property rights are shared for all deserving peoples to enjoy according to their 'ecological civil rights'.

Modern environmentalism is, therefore, about redeploying the institutions of scientific analysis, economic distribution, legal rights and cultural norms in such a way as to ensure fairness of treatment and adequate safeguards for survival for the planet as a whole, including all the people who inhabit it. The planet, of course, will always exist as a marvellously self-regulatory living organism: what cannot be guaranteed is that it will do so in the company of its human species in anything like their entirety.

The challenge of environmentalism for geography teaching

Modern environmental issues cover both discipline and scale in complex and interactive ways. Take the case of stratospheric ozone depletion, for example. There is an interesting 'politics of science' dimension to this. The theory of 'inert gases' having an effect on ozone formation originated in the early days of the artificial creation of the chlorinated and brominated compounds (see Kolowak, 1993 for a useful summary of the science politics of climate warming, ozone depletion and acid rain formation). The major chemical corporations were aware by the mid-1970s of the likelihood of a ban, and hence of the need to create less harmful substitutes. The actual proof depended on a British scientist using traditional methods of painstaking observation to spot the 'hole' in the spring Antarctic. Yet for three years US satellites had measured the discrepancy but had discarded the data on a computer program that systematically removed any 'quirky' evidence that was not within a standard deviation of the norm.

When the 'hole' was discovered two new dimensions to modern science were activated. One was 'megabuck' net-worked research, involving expensive and highly sophisticated measuring technology coupled to fascinating interdisciplinary modelling linking chemists, physicists and meteorologists. The other was 'backlash science' funded by coalitions of vested interests to debunk this 'megabuck science' and then to persuade politicians to turn down the heat on the chemical corporations, at least until the substitutes could be shaped to become commercially viable. Even today that contest is still being waged, with up to 40 per cent of megabuck scientists' time being spent on defending their research record and techniques.

So the science cannot be left to scientists alone to study, because science is deeply embedded in the political structures of finance, expectation and communication. If results are imageable, preferably visually, they command media attention and public interest. The ozone hole, depicted in vivid reds and purples via enhanced computer-modelling techniques, was a classic example. So were the gravitational waves of matter in outer space, a product of the enormously expensive Hubble telescope, that showed that the formation of the galaxies was predictable within the 'big bang' theory. That image alone made the costly Hubble worth while, and would generate the resources to repair it again should it fail to function. And just as images count, so does a good, articulate spokesperson. Science advances nowadays as much by charismatic and fascinating individuals who command the media, as it does in the proverbial 'laboratory'. The politics of global change science very much reflect image, networks, communication and anxiety. These forces shape the institutions that in turn shape the science. Geographers must be aware of this in order to appreciate how an issue becomes framed.

Ozone depletion also becomes of interest because of its spatial configuration. The Antarctic may be a precious wilderness, but it has no vote in international gatherings. So others have to speak on its behalf, and they are never the powerful voices, only the morally concerned ones. The fact that 'sun-avoidance' has become part of the culture of southern Australia, New Zealand and South America, is a defensive reaction, because such peoples again command no special clout. But now that an ozone hole is appearing in the Arctic, and now that there is circumstantial and inferential evidence that ultraviolet light may affect the immune systems of sensitive people, who now stay alive longer because other ailments from which they would normally have died have been eliminated or suppressed, so the geography and politics of ozone removal become *very* different.

The trick for the geographer is to appreciate the different *political* as well as *geographical* scales for analysis, and to recognise that the science of investigation is also influenced by these scales. Then it becomes possible to evaluate the pathways through which the Montreal Protocol (the international agreement for removing ozone-depleting substances) is being negotiated. That in turn requires the knowledge of a reasonable dose of chemistry and international law, *and then* a sense of how the modern transnational corporation is responding to global environmental concerns. As the business community shifts its ethos, so it will have the clout to change national and international agreements and the regulations that underpin them. Always remember that every institutional arrangement is shaped by power, and power in turn is wielded because democracies are assisted to demand wealth creation and commercial solvency as the mainstay of their perceived survival. These are the real politics of ozone depletion: Greenpeace has begun to co-operate with a number of companies, along with the support of the UK Environment Fund designed to promote sustainability technology, in producing ozone-protecting refrigerators that are profitable and competitive. German firms are taking this seriously as are a number of US companies. Inevitably there will be market invasion, but the bigger players will continue to control much of the game with HCFC technology.

A second approach for geographers to follow is that of technique or *research analysis*, and then to look at case studies to reveal the merits or disadvantages of the method. For example, there is nowadays a huge interest in so-called ecologically mediating approaches to environmental decision-making. These include risk assessment or risk management; cost–benefit analysis; ecological economics; life-cycle analysis; ecoauditing; environmental and/or social impact assessment; critical load studies; and the application of the precautionary principle. (For a review, see O'Riordan, 1995.)

This is a heady list and many geographers will still be in the process of familiarising themselves with some of them. Essentially these techniques all have four characteristics:

1. They are multidisciplinary, involving physical measurement and social valuation (or at least judgement) in some sort of organised totality.

2. They seek to quantify and to compare phenomena that heretofore have not been readily measurable and certainly not comparable. They do this by calculating probabilities or weighted outcomes for uncertain or unknown outcomes, and for presuming social biases for the consequences.

3. They are seen as decision tools, aimed at assisting decision-making in public and private life, and in business and government as well as in the household. In practice they carry great influence because the public disquiet about the older 'non-environmental' approaches has reached such an intensity that such techniques are becoming enshrined in the law, in regulation and in decision guidance documents.

4. They cannot be truly effective or 'authentic' unless they are participatory and mediative. This is the tough challenge. All these techniques involve judgements that can no longer be supported by 'expertise' alone. They require a more participatory and democratic grant of reference for their legitimacy. This is sometimes referred to as civic science (Lee, 1993), namely the science of stakeholder involvement, moving sequentially through a series of envisioned possible future states, allowing round tables or other debating and dispute-resolving mechanisms to be created, so as to allow for informed and negotiated consent. In this sense, such techniques are intensely educative phenomena. They require participants to respect and understand each others' viewpoints and needs. They actually alter awareness and therefore political positioning.

Policy piggybacking and the science–politics integration

This last point is the most important for geographers to grasp when contemplating environmental disputes. We are talking about an extended science set in a framework of a fairer and more open democracy, where values and measurements intertwine, and where decision-making structures frame the way in which issues are raised, followed through and interpreted at each stage of their life cycle. This approach is not particularly new. But it is refreshing for geographers who should be best positioned to examine the social consequences of environmental policies, and the environmental opportunities of changing societal concerns. As circumstances shift social perceptions, so environmental issues become captured in wider networks of policy.

Take for example the hugely analysed issue of climate change. Under the UN Framework Convention on Climate Change all of the rich nations of the world are supposed to reduce their greenhouse gas emissions to a level in 2000 equivalent to their 1950 quantities (see O'Riordan and Jager, 1995). How they do this is up to them. As for the EU, the view is accepted, politically and legally, that a number of countries can actually increase their emissions to allow them a reasonable breathing space to develop their economies, while others reduce more to help them out. This, by the way, could be a model for the globe as a whole, to allow the big emitters with money to buy out future greenhouse gas emissions of the would-be greenhouse-gas-rich developing countries, so as to implement jointly a collective reduction strategy.

The fascinating outcome is the envelopment of climate change policies within other policy areas, and the scope for much more of this intermeshing to come. For example, Norway imposed a tough carbon tax, then reduced it in the face of bitter recriminations from the transport sector and the far-flung communities dependent on ferry and air for their goods and services. Subsequently, Norway linked its foreign aid policy to its climate change strategy by financing Polish gas developments and Mexican long-lasting and energy-efficient light bulbs in order to offset CO_2 emissions from outside its borders. Now this trick has backfired, for Norway is in the dock for buying out its CO_2 commitments. Britain imposed a 5 per cent per year petrol 'escalator' tax in order to penalise motor vehicles yet generate much needed revenue (£1.2 billion annually) for the Treasury. Its policy of privatising power and reducing the coal industry to rubble has cost its miners and power workers dear, but cut its CO_2 emissions by 6 per cent. A landfill tax and regulations favouring methane conversion into new energy sources will reduce its methane output by nearly 25 per cent and closure of coal mines also helps the methane strategy. Germany is in the process of removing or improving its coal- or lignite-burning power stations in its new eastern provinces, and in so doing will cut its CO_2 emissions by so much that its western state CO_2 emissions can rise. Emissions in Greece, Portugal, Ireland and Luxembourg can actually rise, yet still the 1990–2000 EU CO_2 envelope will (just) be met.

This is not a success story. Greenhouse gas emissions are falling by default not by design. There is no great political momentum behind climate change precaution and much more difficult decisions have to be faced by all countries in the next few years. So do not be fooled by a misleading piece of political juggling and much preening rhetoric. Yet the message of joint policy 'piggybacking' for environmental gains is one to watch. So too is the significance of national commitments to international agreements and the sophistication of scientific studies and greenhouse gas emission monitoring mechanisms that have begun to be built into the Climate Change Convention process. For better or for worse, a momentum and a supporting verification and enforcement apparatus have been constructed. The combination of a politically extended climate science and a more integrated policy packaging, is a lesson for the future of global environmental negotiations. One should watch for similar developments in the context of the Biodiversity Convention and in the Convention in Combating Diversification. The progression of these developments will be faltering and episodic, but there is still an element of inevitability about it. Geographers would do well to explore the framework of globalisation of environmental performance and commitment through joint implementation and locality studies.

The pathway to a more sustainable economy and society should dominate British politics for decades to come. It may not be always so evident that there is any purposeful shift in policy, technology or social organisation. This is as it should be. Britain will not become more sustainable by design, more by happenstance and opportunism. But various aspects of this transition deserve attention by geographers. One is the move towards a more complete environmental-social audit in parallel with the conventional economic account if from domestic product. This is an argument advanced by the Government's Panel on Sustainable Development (1996). This body urges both a shift to more ecologically-based taxation to release money for socially constructive purposes, and for a more formal green account of the nations' comprehensive progress towards a more environmentally just and socially fair future. This point is picked up by Atkinson and Hamilton (1996) who see a promising future in green accounting of 'unusual' environmental policy areas such as health, education and social care.

Agenda 21, locality and democracy

So far we have noted that environmentalism is capable of endless interpretation, configuration and adaptation to changing economies, societies and technologies. It is this essential adaptability that enables the environmental idea to endure, along with the liberal and caring side of the four tension areas indicated earlier. Environmentalism is already engrained in international and national politics, regulation, business practice and voluntary groups of a thousand different varieties.

Now it is the turn of locality and democracy to embrace environmentalism. The vehicle is local agenda 21 and the device is the linking of the social and environmental agendas. Local agenda 21 is the local government equivalent of Agenda 21, the sustainable development strategy generated at the UN Conference on Environment and Development in Rio de Janeiro in June 1992. The local variety is much more exciting, varied and manageable. It should be the geographer's dream study. It is globally connected, nationally framed but locally shaped and executed. It combines people, work and place in a way that should be unique, yet still essentially part of a global and national commitment. Local agenda 21 may be home grown, very experimental, but still highly connected. It can and should combine the whole of local economy and society—job creation, community rehabilitation, civic responsibility, local economy and environmental health and well-being. It should create the complete notion of security—economic, social and ecological. This, in turn, means open and participatory mechanisms for discussing how a community of whatever size moves from here to there in the sustainability transition. Pilot schemes, experimental projects, linking the schools to local industry, the media and government—all these will be tried and learnt from via electronic data sharing and experience trading, and by perforating the classroom walls of schools, colleges and universities. Local agenda 21 contains within it a concept and a practice of immense power and opportunity. It could and should revolutionise what happens in the classroom, opening up education to the world outside in constructive experiences, twinning experiments and creative use of models, judgements and non-classroom experience. As with the history of environmentalism, geographers can grasp this or let the world pass by. Now is the chance to take action.

What might this mean for geographical education and training in the next millennium?

1. *A better training in the politics of science*, and in the changing role of civic science in the conduct of forecasting; in the application of the precautionary principle and in the creation of mediating and arbitrating techniques, and in the integration of analysis and judgement.
2. *Much closer linkages with local government, business*, regulatory organisations and the voluntary sector by combining research, sharing evidence, creating work placement opportunities and conducting community-based activity in and beyond the classroom.
3. *A greater awareness of who actually gains and loses* nowadays in the pursuit of economic change, and how these individuals and groups will change if any progression towards the sustainability transition is attempted. This is the basis for understanding the distribution of externalities, created inadvertently or deliberately.

This latter focus of activity is a unique zone for the integrationist geographer—connecting the earth and life sciences to the human domain. This should be done, not just by better integration of the disciplines. It also requires envisioning and networking via the community stakeholding groups in imaginative and revealing ways. The task of education is not just to inform; it is also to enable everyone to see things differently and especially from each others' perspective. Local agenda 21 provides the framework for this process. It is up to geographers, along with others, to make it happen.

Note

1. *ENDS* is the monthly report produced by the Environmental Data Services Ltd, Finsbury Business Centre, Finsbury Road, London N5 0NE. A very useful and continuously updated reference on all aspects of environmental policy and regulation.

References

Ashby, E. (1977) *On Reconciling Man with his Environment*. Oxford University Press, Oxford.

Atkinson, G. and Hamilton, K. (1996) Green accounting: monitoring and policy implications. In T. O'Riordan (ed.), *Ecotaxation*. Earthscan, London (in press).

Beckerman, W. (1995) *Small is Stupid*. Jonathan Cape, London.

Bennett, G. (1992) *Dilemmas: Coping with Environmental Problems*. Earthscan, London.

Brown, K. and Pearce D.W. (eds) (1994) *The Causes of Tropical Deforestation*. UCL Press, London.

Brown Weiss, E. (1989) *In Fairness to Future Generations*. Dobbs Ferry Press, New York.

Bullard, R.O. (1994) Overcoming racism in environmental decisionmaking. Environment, **36**(4), 10–20, 39–44.

Cairncross, F. (1995) *Green Ink: Guide to Business and the Environment*. Earthscan, London.

Cameron, T., Werksman, T. and Sands, P. (eds) (1995) *Improving Compliance with International Environmental Law*. Earthscan, London.

DoE (Department of the Environment)(1996) *Second Report of the Government Panel on Sustainable Development*. DoE, London.

Frankhauser, S. (1995) *Valuing Climate Change: the Economics of the Greenhouse Effect*. Earthscan, London.

Haigh, N. and Lanigan, C. (1995) Impact of European Union on UK Environmental policy making. In T.S. Gray (ed.), *UK Environmental Policy in the 1990s*. Macmillan, Basingstoke, pp. 18–37.

Hayes, P. and Smith, K. (eds) (1993) *The Global Greenhouse Regime: Who Pays?* Earthscan, London.

Hays, S.P. (1957) *Conservation and Efficiency: the Progressive Conservation Movement*. Harvard University Press, Cambridge, Mass.

Hays, S.P. (1989) *Beauty, Health and Permanence: the American Conservation Movement 1945–1975*. Oxford University Press, Oxford.

Kolowak, M. (1987) Common threads: research lessons from acid rain, ozone depletion and global warming. *Environment*, **6**(2), 12–17, 29–35.

Lee, K. (1993) *Compass and Gyroscope: Integrating Science and Politics for the Environment*. Island Press, New York.

Leopold, A. (1949) *A Sand County Almanac*. Oxford University Press, Oxford.

Milbrath, L.E. (1989) *Envisioning Sustainable Development*. Suny Press, Buffalo, NY.

North, R. (1995) *Life on a Modern Planet: Rediscovering Faith in Progress*. Manchester University Press, Manchester.

O'Riordan, T. (ed.) (1995) *Environmental Science for Environmental Management*. Longman, Harlow.

O'Riordan, T. and Jager, J. (eds) (1995) *The Politics of Climate Change in Europe*. Routledge, London.

O'Riordan, T., Wood, C. and Shadrake, A. (1993) Interpreting future landscapes. *Journal of Environmental Planning and Management*, 2(1), 37–47.

Pearson, N. and Smith, S. (1991) *The European Carbon Tax: an Assessment of the Commission's Proposals*. Institute of Fiscal Studies, London.

Sands, P. (ed.) (1994) *Greening International Law*. Earthscan, London.

Turner, R.K., Pearce, D. and Bateman, I. (1993) *Environmental Economics: and Elementary Introduction*. Harvester Wheatsheaf, Hemel Hempstead.

Part C

Geography in the school curriculum

Geography 5–19: retrospect and prospect

9

Rex Walford

Introduction

Now that the National Curriculum tumult has ceased (more or less) and the curriculum captains and kings (Kenneth Baker, Duncan Graham, John Patten et al.) have departed for pastures new, it is an opportune time to consider the state of geographical education and its prospects for the twenty-first century. The past few years in England and Wales have been so filled with the hurly-burly of government action and its ramifications that there has been little chance to assess the state of play in anything other than a fleeting way. One year's apparent certainties have turned into the waste paper of the next; that which was praised by one Minister was frequently overturned by a successor. It is worth recording that between the idea of the National Curriculum being mooted in 1987 and its 'final revision' in 1995, there have been no less than five Secretaries of State for Education (Baker, MacGregor, Clarke, Patten, Shephard) defining policy in their various ways.

A revision was instituted by John Patten in April 1993, in the light of much teacher complaint about overload and disquiet about problems of assessment (Daugherty, 1995). If handled with more tact the grumbling might have been quelled without further wholesale change, but in the end a full review of National Curriculum matters was set in train

Geography into the Twenty-first Century, Edited by E.M. Rawling and R.A. Daugherty,
© 1996 John Wiley & Sons Ltd.

under Sir Ron Dearing. The 'Dearing Review' (Dearing, 1994) spawned another set of Statutory Orders and a promise that 'all will now be left alone for the next five years'. For this relief, much thanks. . . .

What place for geography in the twenty-first century curriculum?

The review of the National Curriculum instituted by Dearing received a good press in most areas and clearly alleviated some of the general problems of overload within subjects and in the curriculum as a whole. It clarified and improved the original Geography Order in some respects though the responses of geographers were initially cautious and wary rather than enthusiastic (Bennetts, 1994; Carter, 1994). But (as is explained below) there was reason for geographers not to share fully in the general euphoria which greeted the proposals.

Though Kenneth Baker's National Curriculum plans of 1987 owed more to intuition than to discourse with educationists (see the disappointment of theorists like John White and Paul Hirst (White, 1988; Hirst, 1993)), there was little general dissent from his intention to continue to operate on a traditional basis of 'subjects'. This was already the *de facto* situation in the overwhelming majority of secondary schools, despite a patina of discourse about 'areas of experience' which was fashionable in HMI circles for a time. It has also proved surprisingly acceptable to primary schools, following a quarter of a century in which the 'integrated day' (encouraged by the 1968 Plowden Report) has had time to be explored and evaluated.

However, also at the heart of the Baker curriculum plan was a desire to see a broad and balanced curriculum based wider than English, maths and science. The latter formulation was advocated by the then Prime Minister, Margaret Thatcher, but, as Baker's political memoirs clearly show (Baker, 1993), this was a rare occasion in which her will did not prevail in Cabinet. Baker argued successfully for a National Curriculum formulation to include history, geography and the arts for all pupils up the age of 16—a notable expansive vision which was rarely acknowledged in the later turbulence surrounding his proposals.

It was subsequent reformulations which began to pare

away at this, culminating in John Patten's inadequate management of relationships with teachers which led to the Dearing Review. Dearing (with no previous experience in educational matters other than as a school pupil) acted as arbitrator and 'honest broker' in trying to patch up the situation. In finding a pragmatic solution he effectively diluted the concept of the 'broad and balanced curriculum' by according more emphasis and status to the Thatcherian core (English, maths and science) and by making half of the National Curriculum subjects (including geography) optional after the age of 14. The alternative concept of a 'tiered curriculum' is reinforced by the pattern of national assessment testing and in the directed emphases for inspections of schools.

Dearing's resolute avoidance of the curriculum implications is clear to see in the report he wrote. 'History and geography are absorbing and valuable subjects. But I cannot see a reason . . . why these subjects should . . . be given key priority in this key stage over others' (Dearing, 1994, p. 46). For Key Stage 4 the question of breadth, balance and the relative merit and complementarity of subjects was thus jettisoned in a context of disarming pragmatism. Geography has benefited from a restored visibility in primary schools as the result of the National Curriculum debates, and in secondary schools its position is more secure as a separate field of study in the 11–14 age range than it was before the Education Reform Act of 1988. It is in the upper age ranges that its position faces future erosion.

The Dearing Review also introduced the concept of alternative curriculum pathways for students in technical and vocational subjects from 14 years onwards, a move disliked by many parents, as the results of the subsequent consultation exercise showed (SCAA, 1994a). The protestations that these alternative pathways will have 'parity of esteem' have an uncomfortable echo from the experience of setting up grammar, technical and secondary modern schools from the 1944 Education Act. The pathways are being hindered from the outset by the somewhat chequered reputation which new qualifications such as GNVQs are already acquiring. Perhaps more critically (though not necessarily justifiably), if these qualifications fail to gain the confidence of admissions tutors in institutions of higher education, they are likely to be deemed second class to the traditional 'gold standard' of A levels for the foreseeable future. So their status remains, at present, problematic.

The take-up of GNVQ courses is, however, growing steadily. Some geography teachers are optimistic that the present reservations about them will disappear, and support the concept of the alternative pathways for general educational reasons. They see opportunity in the fact that there are elements of geography in some of the GNVQs already set up—notably that in leisure and tourism. But though this may provide employment for existing geography staff in institutions where the new qualifications are being introduced, the experience for students will only include a fragment of geography. It will be quite different from the holism and the synthesis of physical and human geography which will characterise GCSE and A level study in the subject from 1996 onwards.

Geography's place in the 14–16 curriculum would be secure even if optional, if there were free choice of subjects; it has proved itself consistently popular in GCSE in the past (see Appendix 3). However, a diversion of students to vocational and technical pathways is almost sure to diminish the opportunity to study geography to GCSE, if not the desire. And this subsequently will have knock-on effects. Geography's currently healthy position in A level statistics (see Appendix 3) reflects a steady growth in numbers over the last few years, but the plateauing effects in the candidature are already beginning to show. A future decrease in the share of GCSE numbers will be likely to reduce the A level entry—and, in turn, reduce the pool of good candidates who wish to read the subject at university. Thus, decisions made in one part of the education sector have effects throughout the whole.

As the twenty-first century begins, geography may well be looking with some concern at a year-by-year decrease in the numbers who study it as a whole at all levels of education. The bright, hopeful days of the late 1980s following the occasion when Kenneth Baker welcomed a GA delegation to his office and told them of his intention to give geography (along with history) a secure and substantial place in the 5–16 curriculum will be a wistful memory. In the future, there will be stiff challenges to maintain the status of 'foundation' subjects against a re-emphasised core curriculum in the 5–14 age range, and to attract and recruit students in the crowded market-place of option choices at 14, 16 and 18.

What kind of geography?

One of the uncomfortable realisations engendered by the intervention of the laity into the Reform Act education debate (and the subsequent deliberations) is that innovations acclaimed within the profession are sometimes critically regarded from beyond it. Thus, accorded some access through their membership of curriculum working groups and through the general consultation process, a new set of voices have begun to make themselves heard. The government has encouraged a form of 'curriculum democracy' which gives heed to the views of parents, employers, and governors of schools, as well as those of teachers, teacher-educators, HMI and curriculum developers. Sometimes such lay voices offer only half-formed views or marginal comment; but it is unwise to ignore or dismiss them, since they represent the constituency which ultimately receives the fruits (and pays the bills) of education.

In geography, a group of five such voices was heard in the original Geography Working Group which produced the original National Curriculum plan (DES, 1990), the basis of the first Statutory Orders (DES, 1991; WO, 1991). In hindsight, it can be seen that they made a discernible impact on the thinking of the Group, and on the nature of the geography which was recommended for the 5–16 curriculum. This situation was powerfully reinforced by the fact that the DES (indirectly and directly) made it clear that the current situation in geography education was not something which they wished to be sanctified.

Much of the fresh basic framework suggested by the Group remains in the revised version of the National Curriculum (DfE, 1995; WO, 1995), despite changes to its form and style in the period since. Concern about lack of general world knowledge (evidenced by a number of well-publicised international surveys in the late 1980s) resulted in a decision to require some specific locational knowledge, as well as a renewed emphasis on the study of actual places at a variety of scales. Identifying places to be learnt on maps of the UK, Europe and the world proved a much less contentious issue than was at first feared, and proved a useful flag to signal intentions. (It is interesting to speculate whether history might have avoided such a rough ride had it made a similar decision.) Even after the Dearing Review, the pages of maps with required locations survive;

the revising group and SCAA officials concerned themselves with reallocating place knowledge from the attainment targets to the programme of study and with the minutiae of which places might or might not be legitimately required, rather than discarding this 'extra' requirement.

But, in truth, this is an issue of intermittent public interest rather than great educational significance. More central to the proposals was a concern to ensure a return to a more secure base of knowledge, given some of the apparent effects of curriculum experiments in the 1970s and 1980s. It was a strategy undermining the belief that curriculum development always pushed the subject onwards and upwards and owed more to Housman's circumspect view of 'progress': 'Until it starts, you never know in which direction it will go.'

A return to a focus on the study of places, an affirmation that physical geography deserved as much attention as human geography, and a new emphasis on issues concerning the management and stewardship of the environment were keystone ideas in the original discussions—and have survived the transition through from an original seven attainment targets to the final version of a single one (Rawling and Westaway, 1994). The final post-Dearing version of the geography curriculum, though much slimmed down and changed from the 1990 one, still bears the recognisable frame of these ideas, and current discussions among teachers are about their translation and interpretation rather than their fundamental legitimacy. The 5–14 geography curriculum of the early twenty-first century in schools is likely to bear this hallmark.

In practice this will mean a return to more place-based case studies in both primary and secondary schools and (one suspects) a rediscovery for all of the innocent pleasure of frequently poring over globes, atlases and maps. The concern for students to learn more geographical knowledge will be found not to necessarily imply a concentration on the acquisition of trivia. Teachers will rediscover an enjoyment of parts of physical geography which they had long neglected and gradually develop a thorough understanding of contemporary environmental issues. The impetus to change engendered by the original Geography Orders has already produced many new units of work, and the impulse will probably be to select the most appropriate and then refine and improve these rather than embark on yet another set of radical changes.

At GCSE the introduction of a new set of syllabuses from 1996 onwards will be undertaken with a new set of criteria. The syllabuses will bear the imprint of that part of the geography National Curriculum that never was (14–16), though now, of course, they will only be taken by a percentage of the school population. One precious aspect of the new syllabuses is that every student is required to undertake fieldwork in the pursuit of coursework investigations; so that the last cohorts of 'armchair geographers' in the teaching force will find it more and more difficult to maintain a totally factual and/or theoretical approach. The revised GCSE geography criteria represent a common core for the new breed of syllabuses and the input of (originally) SEAC and (currently) SCAA has been considerable in these areas.

At A level, new syllabuses are also being introduced (from 1995). The shift of interest from micro- to macro-scale matters in the world is reflected by the greater emphasis on global issues in these, though case studies remain an important element. Here also, the balancing of physical geography with human becomes apparent in the core requirements, and in the specification of an integrating study. Perhaps most significant may be the increased take-up of modular courses in which the qualification can be achieved (and improved) in steps rather than at a single sitting.

The new National Curriculum order (5–14), the revised GCSE criteria and the new A level criteria and core offer, for the first time, a spine of common requirements for those who choose to take the subject through to the age of 18.

The tide of statutory regulation has not yet lapped the shores of higher academe. However, geographers in universities may need to keep a vigilant outlook if they wish to retain their independence in curriculum matters, given the speed with which the basic idea of a National Curriculum for schools has become accepted wisdom. The current disparity in what different institutions of higher education offer by way of a degree in geography cannot be a help in this situation; it is quite possible, as teacher-educators well know (Bale and McPartland, 1986), to be confronted by two aspiring teachers with geography degrees who have practically nothing in common with each other in terms of what they have studied.

There are good educational reasons why some restraint might be exercised in the smorgasbord of courses and topics available in some places. Even given the views of those who glory in its postmodernist tendencies (see Jackson, Chapter 6) there is a *political* need for the subject itself to seek some coherence and to reassert traditional values of holism and synthesis. To do so does not minimise the value of the insights of those on the frontiers of the subject, nor require that a single monolithic philosophy of the subject be held.

Professor Alice Garnett, a long-time servant of educational causes within all levels of geography, addressed herself to the question 'Whither the geography of today and tomorrow?' in her last published words (Garnett, 1987, p. 241). Her answer was both acute and salutary . . .

> [It] will depend [on] decision-making regarding the educational value and popularity of our subject in schools and its role as a university discipline. . . . As geographers today probe and strive to expand our frontiers further and further, would our answers now be too disparate to command attention.

What kind of pedagogy?

One recurring fear expressed about the original National Curriculum proposals in geography was that the amount of material specified would lead to heavy didacticism; in the effort to cover all that was required, teachers would resort to dictating notes, return to requiring learning by rote, avoid participatory methods because they slowed up the transmission process. Even though the government protested that prescription of pedagogy was not on their agenda, the consequences of 'overload' were predicted to be adverse in this area.

Subsequent reformulations have gone a long way to putting this right, and the revised Dearing formula encourages 'an enquiring, questioning approach', and makes a commitment to the continuation of fieldwork. Whether this works out in practice remains to be seen, but there is every hope that schoolteachers will continue to use a variety of methods in geography classrooms, and maintain a long and honourable tradition of pedagogic enlightenment in geography which stretches back more than a century.

Even in the Victorian schoolroom, it was geography which was most conspicuous by its use of teaching aids (Baigent, 1994) including maps and artefacts. The original founding of the Geographical Association in 1893 came about through an intention to form a group for the exchange of lantern slides (Balchin, 1993) and in the first half of the twentieth century many geography teachers followed this up with an enthusiasm for the use of film, radio broadcasts and other practical resources (Boardman and McPartland, 1993). More recently, the advent of audiotape and videotape has increased flexibility in the use of sounds and images.

Now, as is recognised almost universally, we stand on the edge of yet another major communications revolution and the potential of CD-ROM, interactive video, electronic-mail networks and other new computer-linked aids wait to be harnessed. There are plenty of encouraging indications (such as the recent co-operation between the GA and NCET in providing support for IT through conferences, INSET and teaching materials) that geographers will remain in the thick of these developments, though some will inevitably embrace them more enthusiastically than others. But the very richness of image and variety of action available through these new developments create an irresistible magnetism for those who love and cherish their subject, as long as they have not fallen prey to a minimalist approach to their teaching.

In higher education, the same factors apply, with the greater possibility of being involved in the research and development of particular approaches. The rise in class sizes in higher education (Jenkins and Smith, 1993) may, paradoxically, be one of the most pressing reasons for the introduction of new technologies into teaching and the move away from traditional large lecture situations.

But alongside the potential of contemporary modes of information and educational technology in schools, there is also a more worrying tendency in relation to resource provision. With many schools given greater autonomy over spending decisions, the desire to maximise efficiency and cut costs in a time of relative stringency can result in pressure not to be profligate (Kington, 1994). In particular, the phenomenon of the 'single course book' is now encountered more frequently in geography classrooms at all levels. In spite of the undoubted efficiency and excellence of some texts carefully tailored for their market, the dangers of the single source for information are nevertheless considerable. It would be unfortunate if, mistakenly secure with a book which appears to do everything, hard-pressed geographers were to give up their usual search for alternative formulations and viewpoints and cease to comb current newspapers and magazines for up-to-date materials.

Another potential difficulty arises from the changes in teacher education which are being promoted by the Department for Education and Employment. Current ideology suggests that a shift to school-centred (SCITT) schemes should be promoted, at the expense of schemes which involve partnerships between schools and institutions of higher education. There appears, as yet, no great enthusiasm for SCITT schemes in more than a handful of schools but the inducements to go down this path may be increased in the future in England by the newly formed Teacher Training Agency.

A resultant effect of such a change may be to alter the nature of training for new teachers, with an increased pressure to conform to existing practices in the training school, rather than to experiment more widely. In SCITT schemes, students will be likely to have less time for specialist subject studies, less opportunity to mix with other students in their subject, less contact with experienced specialist teacher-educators, and less chance to visit a variety of schools. The consequences for geography, as for other subjects, could be to slow down innovation and development, and to promote a culture of conservatism and restrictive dependence in new teachers.

But overall, pedagogy in schools in the early twenty-first century looks likely to be driven by renewed consciousness and concern about assessment. (The National Curriculum plan was, after all, assessment driven in the first place.) This is likely to engender fresh emphasis on purpose and objectives, but (in geography at least) not necessarily to lead to didacticism because of the rich vein of resource materials waiting to be tapped. Young minds of the present day do not take easily to long spells of passive learning and geography is better placed than many subjects to provide stimulation and enrichment of experience on the one hand, while maintaining purposefulness and relevance to contemporary life on the other.

Conclusion

The modifications made by Sir Ron Dearing's review of the National Curriculum have been welcomed in many quarters, but are not advantageous overall to geography. Their best feature is that they add clarity to the curriculum and reduce it to manageable proportions and promise a period of calm and stability for the next few years. Within that, geographers in both primary and secondary schools may be able to settle more systematically to re-establishing a balanced geography (of places, themes and skills, graphically represented in the original 'cube' diagram developed by the Geography Working Group) and to advance new developments in practical ways. The moratorium on major change may also help teachers return to the mode in which considered reading and reflection about the *content* of geography was a normal part of preparatory activity.

But the pressures of competing elements in the school curriculum post-14 will require secondary-school geographers to be active in epitomising and extolling the virtues of their subject. The need to 'sell' geography (Kent, 1990) at parents' meetings, option evenings and open days will be paramount, and the assistance of national events (such as the annual Careers for Geographers symposium organised by the RGS and the Geography Action Week being planned by the GA for 1996) will become an important extra factor to be used to advantage.

Geographers in higher education will need to help nurture their own seed-corn if they are to continue to teach students of quality in geography degrees; perhaps, renewed interest in the GA, and, through COBRIG, in the whole spectrum of geographical education may emerge. In a previous era, it was common practice for professors and lecturers of geography in universities to stump the country speaking to GA branches, WEA conferences, and school lecture occasions in order to propagate their belief in the virtue and practicality of their subject. Mackinder, Herbertson and Wooldridge were prime examples of those who covered many miles and spent many hours in the pursuit of this wider objective. They did so because they knew that they needed to acquire adherents and disciples to the geographic vision. Today, the insistent demands of publication and research make it more difficult to allocate time to such activity. But there may be no research to pursue if the supply of students dries up.

There is also a need for geographers in higher education to engage in renewed debate about, and to find a commitment to, an acknowledged 'core' to the subject to prevent excessive fragmentation of activity and of the whole subject image. If not, the future may be bleak for geography degree enrolments, in the longer term.

Geographical education will enter the twenty-first century with important issues to be faced at all levels; the traditional vigour and enthusiasm of those who teach the subject will need to be matched by careful strategic planning and by alert opportunism in order to meet the challenges.

References

Baigent, E. (1994) Recreating our past; geography and the rewriting of the Dictionary of National Biography. *Transactions of the Institute of British Geographers*, **19**(2), 225–227.

Baker, K. (1993) *The Turbulent Years; My Life in Politics*. Faber and Faber, London.

Balchin, W.G.V. (1993) *The Geographical Association: the First Hundred Years*. Geographical Association, Sheffield.

Bale, J. and McPartland, M. (1986) Johnstonian anarchy; inspectorial interest and the undergraduate education of PGCE students. *Journal of Geography in Higher Education*, **10**(1), 61–70.

Bennetts, T.H. (1994) The Dearing Report and its implications for geography. *Teaching Geography*, **19**(2), 60–64.

Boardman, D. and McPartland, M. (1993) A hundred years of geography teaching. *Teaching Geography*, **18**(1), 3–6.

Carter, R. (1994) Feet back on firmer ground. *Times Educational Supplement*,18 Nov. 1994, p. III of Geography Extra.

Daugherty, R. (1995) *National Curriculum Assessment: a Review of Policy 1987 to 1994*. Falmer, London.

Dearing, R. (1994) *The National Curriculum and its Assessment; a Final Report*. SCAA, London.

DES (Department of Education and Science) (1990) *Geography for Ages 5–16; Final Report of the Geography Working Group*. HMSO, London.

DES (1991) *Geography in the National Curriculum: Final Orders*. HMSO, London.

DfE (Department for Education) (1995) *Geography in the National Curriculum*. HMSO, London.

Garnett, A. (1987) Comment: 'The Pioneers'; some recollections. *Transactions of the Institute of British Geographers*, **12**(2), 240–241.

Hirst, P. (1993) The foundations of the National Curriculum: why subjects? In P. O'Hear and J. White (eds) *Assessing the National Curriculum*. Paul Chapman, London, pp. 31–37.

Jenkins, A. and Smith, P. (1993) Expansion, efficiency and teaching quality: the experience of British geography departments 1986–91. *Transactions of the Institute of British Geographers*, **18**(4), 500–515.

Kent, A. (1990) *Selling Geography*. Geographical Association, Sheffield.

Kington, C. (1994) Resourcing the National Curriculum. In R. Walford and P. Machon (eds), *Challenging Times; Implementing the National Curriculum in Geography*. Cambridge Publishing Services, Cambridge.

Rawling, E. and Westaway, J. (1994) Choose your destination. *Times Educational Supplement*, 18 Nov. 1994, p. II of Geography Extra.

SCAA (School Curriculum and Assessment Authority) (1994a) *The Review of the National Curriculum; a Report on the 1994 Consultation*. SCAA, London.

SCAA (1994b) *The National Curriculum Orders*. HMSO, London.

WO (Welsh Office) (1991) *Geography in the National Curriculum (Wales)*. HMSO, Cardiff.

WO (1995) *Geography in the National Curriculum (Wales)*. HMSO, Cardiff.

White, J. (1988) An unconstitutional National Curriculum. In D. Lawton and C. Chitty (eds), *The National Curriculum*. Bedford Way Papers No. 33, London Institute of Education, London.

10 Developments at A level

David Hall

The background to change

Growth and development

Advanced levels have existed for almost half a century. It is sobering to think that I sat the last examination (in 1950) of its predecessor, the Higher School Certificate, with its pink, green and yellow tinted papers for subsidiary, principal and scholarship levels. The first main paper in geography comprised one compulsory Ordnance Survey map and three other questions from which I chose one on the method of construction of the polar zenithal equidistant and Sansom Flamsteed projections, one on the chief producing areas and mining methods of copper on a global scale, and one on the physical and economic characteristics of plantation agriculture in the tropics. Paper 2 covered the other face of geography with sections on Western Europe/British Isles; the Mediterranean region, etc.; the monsoon lands of Asia, etc.; and the Americas/Australia, etc.; and the 'etceteras' covered between them the former USSR, French North Africa, the Belgian Congo and the Middle East. There could presumably be no grounds for litigation that any one major region of the globe had been omitted. I recall giving a reasoned account of the main regional variations in the density of population of China; probably a Roxby favourite attributable to the influence

Geography into the Twenty-first Century, Edited by E.M. Rawling and R.A. Daugherty,
© 1996 John Wiley & Sons Ltd.

of the Liverpool Geography Department on the Northern Universities Joint Board.

In those days competition for means rather than competition for places was the determinant of access to a university education, and in this respect the yellow scholarship papers had a crucial function in the aggregation of total marks for state or county scholarships. Today, scholarship papers are just a symptom of the 'sabre toothed curriculum' (Benjamin, 1939), but even a single grade change by one letter in the core papers in a subject can destroy an intended career. The importance of grade points for university admission places an immense responsibility on the A level system to offer examinations that are fair, reliable and valid. Yet the scholarship essay questions set in 1950, with their challenge to literacy in that free response mode—'Discuss the view that Canada is the America of Opportunity, the United States the America of Achievement'—can still be found in the papers of some subject syllabuses today.

The growth of geography in higher education since the First World War, and of the part played by key individuals in subject departments in both Oxbridge and redbrick institutions in this process, is a familiar story. The interests of a number of those individuals extended beyond the frontiers of academia to the encouragement of geography in schools, not only by their support and activity within the Geographical Association (GA), but also through their involvement with the construction of syllabuses and the setting and marking of papers at HSC/GCE A level. One motivating factor undoubtedly was a sense that geography was an essential part of a liberal education for the young person, but another was the political awareness that the best guarantee of a lively and knowledgeable undergraduate intake rested crucially upon the subject learning and personal interest fostered by teaching of high quality in the schools. Throughout the post-war period, geography continued to attract a significant number and a growing proportion of the total entries. During the 1960s it overtook French and chemistry, and, although itself overtaken by Economics, by 1971 it had become the sixth most popular subject with 22 000 candidates, an increase of 140 per cent over 1961. By 1994 it ranked fourth among the main school subjects with an entry of over 46 000 candidates (see Appendix 3).

Doubtless any aspiring Managing Director for Geography (UK) plc would provide a glowing preface to the Annual

Report congratulating all the staff for their contribution to this result. Analysts would point out that for 20 years there had been a growth market sourced by the rise in the participation rate of the 16–18 age group in full-time education. They would attribute the increase in market share taken by geography to the success of the Bristol 14–18 and the Avery Hill Projects throughout the 1980s in preparing the foundations for A level choice, and the appearance, through Geography 16–19, of a new A level product ideally suited to the needs of today's post-16 students, with an emphasis upon contemporary environmental and spatial issues. Yet the sudden and unexpected decline of economics is a reminder that past performance is no guarantee of future success. Of course this can be explained away as a particular difficulty of the 'dismal science' when faced with the competition of business studies, but the appearance of GNVQs as an alternative route post-16 gives a sense of uncertainty about the future of A level in all subjects and adds weight to the views of the bears in the current market-place of student choice.

The GCE examination boards

With the closure of the Southern Universities Joint Board from 1990, eight independent examining boards remain in England, Wales and Northern Ireland, all offering geography as an A level subject examination. Figure 10.1 shows the dominance until the 1990s of the Joint Matriculation Board (now NEAB—the Northern Examinations and Assessment Board) with its large number of centres in the North and the Midlands. The University of London Schools Examination Board (ULSEB) was for many years the second major provider with its Syllabus 9210, which was first offered in 1978 following substantial changes to the pre-existing syllabus, explaining perhaps the temporary fall in the candidature at that time.

The examination set by the University of Oxford Delegacy for Local Examinations (UODLE) was seen by many teachers throughout this period as being conservative in its approach, with an emphasis upon clearly expressed communication in essays of traditional length (three in two plus hours) on physical, human and regional topics. By contrast the University of Cambridge Local Examinations Syndicate

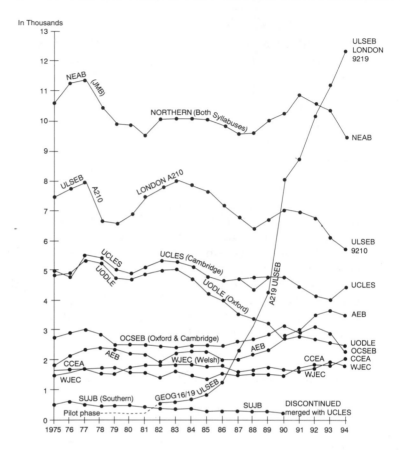

Figure 10.1 Candidates sitting A level examinations in geography by exam board, 1975–94

(UCLES) seems to have retained its client base. Its questions typically carried a strong scientific call for data and evidence rather than an emphasis on elaborated factual descriptions or case studies, as might be expected from the questions set by the Oxford Delegacy or even the London Board (A9210).

OCSEB, the Oxford and Cambridge Schools Examination Board, with its clientele drawn mainly from the independent sector, introduced a fundamentally new syllabus in the 1970s in which a systems approach was stated as the central unifying theme and as the criterion for relevance in the assessment of quality. The Associated Examining Board (AEB), the largest board in terms of the total number of entries across all subjects, has always enjoyed a considerable following in the further education sector. During recent years efforts have been made by the AEB to raise numbers in geography to levels comparable with other major subjects, with a distinctive syllabus that had a holistic approach to geography as 'a study of Man and his behaviour in ecological and spatial contexts'.

With around 2000 candidates each, the Welsh Joint Education Committee (WJEC) and the Northern Ireland Board which is now part of the Council for the Curriculum, Examinations and Assessment (CCEA), serve distinct communities, with WJEC setting its examination papers and marking scripts in Welsh as well as English. In the 1980s, their approach to syllabus content and to assessment design was innovative and will be considered below.

But, while each of the boards has sought to maintain and develop its distinctive approach to the subject at A level, the most notable of all changes shown by Figure 10.1 is the rise to pre-eminence of Geography 16–19 A level. The Schools Council funded this project for five years from 1976. ULSEB accepted the task of developing the scheme of assessment with the project team and then adopted the examination as the A219 syllabus in the early 1980s when the examination was thrown open to all centres at the end of its pilot stage. Its rapid growth affected the numbers reported by all other syllabuses. Geography 16–19 also created a new market by attracting students from other subjects, and thereby encouraged students to take up the challenge to A level. It was a syllabus that had particular appeal to many 16–19 students in the 1980s and early 1990s, and, as shown in Figure 10.1, by 1994 it had become the largest of all A level geography syllabuses with 27.6 per cent of the total entry. ULSEB thereby became responsible for 40 per cent of all A level subject entries. The reasons for this exceptional growth and its substantial impact on A level geography will be examined further below.

Assessment standards at A level

For many years the matter of grades at A level was contentious. Originally the five pass levels A–E were norm-referenced candidate percentages of 10:15:10:15 from A to D with 20 per cent allocated to E. However, this created a narrow band at grade C where candidates tended to be bunched together, with only a few percentage points to separate a grade B from a grade D. Since the mid-1980s the central block of candidates between the B/C and the D/E boundaries have been divided arithmetically by trisecting the interval of the mark range in equal proportions and not dividing the grades according to candidate percentiles. Theoretically, the established norms of the standards reported by the A and B grades have been unaffected, and the number of C grades has been increased. Since 1987 the 'O' (equivalent to a GCE pass) and 'F' (fail) grades have been replaced by 'N' (near miss) and 'U' (unclassified) which are arithmetically divided equally below the E/N boundary.

In the early days of A level, general agreement about quality was implicit, in a shared understanding between the lecturers in university who were the chief examiners setting and marking the papers, and the teachers in the schools who were themselves graduates and happy to work as junior partners in what Reid (1978) has called the 'university connection'. Syllabuses were short lists of content, which required little elaboration as the established texts such as Monkhouse could be consulted as necessary. Timed essays, a map question and some cartographic representation in a practical paper provided all the evidence necessary to place students in rank order. Mark schemes were merely prompts set by chief examiners to kick start the marking process, and were confidential to the team of assistants. This permitted experienced lecturers, prepared to do the thankless task of marking at the hottest time of the year, the power to exercise flexibility in judgements and to mark holistically, thus preventing the achievement of higher grades by a process of mark grubbing on a point-by-point basis.

In this 'broad view' of assessment, it did not matter that essay questions had different levels of demand, or that options provided unequal routes through an examination. In any answer the criteria outlined in the Robbins Report (1963), based on the quality of constructing and communicating a reasoned argument in prose, could be applied. Essay papers performed this function, and a judgement about quality was made on a question-by-question basis. The 'best' candidates were those able to combine the shorthand of nomothetic terminology pertinent to the theme, with an information base appropriately selected and emphasised through a conceptual understanding of the subject matter. Variation in the level of demand made by particular questions was of lesser importance as the most able candidates would select the most challenging questions, and those with little idea of their subtleties would display their incompetence. In addition, an invisible hand existed to see that justice be done, as a trade-off could be expected between those questions where a candidate might display an apparent weakness with those where she/he might display apparent strength. Of course, for this to happen a critical number of questions of similar type had to be answered and candidates were reminded of the importance of answering fully the number of questions specified in the rubric. At the 'award' stage, responses to questions of high cognitive demand could be selected and reviewed as threshold indicators for candidates at or near a grade boundary, and the buds, if not blooms, of incipient talent could be detected and rewarded.

Since the 1970s this view has been in retreat, as the subject has become more eclectic in its outlook and approach. The growth and change in the composition of the client group, together with the proliferation of choice in textbooks and other sources of information, added pressure to the demand for a more open approach to examining. The examination was no longer a secret garden where the separation of assessment from teaching and learning was held to be the main guarantee of impartiality. At a different level, the Certificate of Secondary Education led the way by introducing syllabuses with content more fully elaborated, explicit statements of aims and objectives and a much greater range of assessment methods. Linked with these developments in the technicalities of assessment was the development of school-based ('mode 3') examinations where schools were encouraged to determine the syllabus, set the examination papers and then mark them, with the boards responsible only for the moderation and awarding procedures.

In consequence, the 'narrow view' of assessment has become the dominant ideology, mirroring the development of theories of management in other social and economic spheres. This view is suspicious of shared consensus about quality. It holds that, since geography is multivariate, assessment based upon essays plus a practical and map question is too restricted a base from which to construct a profile of excellence across the spectrum of knowledge, skills and understanding now required for both summative and diagnostic purposes. It follows that, in view of the wider range of teaching quality, of teaching institutions, and of

Broad	Narrow
Aims/objectives Implicit based on generally accepted collegiate view of the subject and its modes of communication established over time between committed specialists. Any overview offered is an explicandum of aspects of syllabus structure	Explicitly defined distinction between aims as general statements of intent and objectives as statements of performance which can be assessed with some precision. Assessment objectives grouped under the key cross-curricular operators such as concepts, understanding skills, values (the elements of learning)
Syllabus Brief notes of content only to be amplified by reference to core textbooks suitably updated and conventionally listed in a reading list	Considerable amplification often in double column format, one of which lists the key ideas/concepts/processes which will be assessed (the power generators) and the other which offers further amplification or illustration
Examinations Mainly end-of-course timed essays with some additional papers/sections testing specific cartographic/map skills, and possibly externally examined individual studies	An array of assessment instruments suitable for the purposes specified in the aims/objectives: long/short free response, stimulus/data response, short closed response, role play. Coursework, often internally assessed with explicit moderation procedures in place
Question papers Considerable variation in level of demand in essay questions sourced by inspiration only lightly revised for content validity. Question choice within sections of papers with a mixture of open or closed questions	A declared policy to achieve equality of demand between questions where choice allows different routes to be taken by candidates. Use of compulsory sections or even papers (such as decision-making) differentiating by outcome
Criteria of assessment Impression marked on a question-by-question basis with general criteria usually in mind, and often élitist norms based on the top-down idea of the perfect candidate	Attempts to specify performance thresholds by describing a number of levels of achievement for specific components or sections of a paper in the mark scheme
Mark schemes Confidential. Light brush approach. A professional and experienced examining team determine the mark awarded in relation to the quality of response judged to have been achieved in each particular essay assessed	Explicitly formulated in some detail with marks and submarks distributed with reference to the weightings allocated by the syllabus objectives for the component being assessed, interlocked with the level descriptors
Standards Maintained by norm referencing grade percentages and allowing performance to float. Validity underwritten by range and scope of a candidate's response aggregated across the whole examination	Maintained by level descriptors which rank candidates on the basis of their performance component by component rather than upon aggregation and norm referencing the final candidates' list

Figure 10.2 Broad and narrow approaches to assessment

possible future routeways into higher education, syllabuses should be issued which give, to teachers, specific guidance for the preparation of their schemes of work, and to students, an assurance that they understand explicitly during their course of study, what is being assessed. Figure 10.2 is a summary comparison of these viewpoints. I have discussed elsewhere in more detail (Hall, 1991, 1994), the dynamics of the broader curricular context in which this shift has occurred, and the related cultural perspective of an enterprise, rather than an ecological, culture in which it is located (Hall, 1990).

The Geography 16–19 framework and its consequences

By taking an overall viewpoint about the kind of geography which would best meet the needs and interests of the 16–19 student, and constructing a curriculum framework, a route for enquiry, and then an assessment system designed to support those aims, Geography 16–19 avoided the mechanistic dangers of the narrow view and provided a template for development at A level which was both explicit and holistic. It thereby made an immeasurable contribution to the planning of A level models of curriculum and assessment. It built upon the achievements of the Bristol Geography Project, which had recognised the importance of the examination boards as the gatekeepers of the 14–16 age teaching/learning process in schools, and which had negotiated with the UCLES the idea of enhanced professionalism of teachers, by their participation in assessment through a combination of coursework and moderation procedures. It had access also to work done through the Schools Council and the Geographical Association (GA), by working parties producing new ideas on A-level syllabuses and examination procedures related to the five subject, two level N & F proposals.

Through preliminary conferences and its steering committee consultants, evidence of the 'orchestrated pressures for change' (Naish et al., 1987) was collected. In consequence, the broad aims of geography were identified in terms of the potential of the subject in contributing to the needs of a 16–19 client group, broadly defined beyond the narrower focus of preparation for a degree course in the subject field. Feedback from questionnaires sent out nation-wide indicated that teachers wanted to teach in more active mode using environmental themes of direct significance for the pupils they taught. They resented the pressures of overloaded syllabuses stuffed with learning prescriptions drawn from the shelf knowledge of textbooks whose pages were yellowed by the passage of time, with dictated notes accepted by many teachers as the only way that course requirements could be fulfilled. To meet these pressures the Project devised a framework as the key for its proposals.

Content was seen not as a collection code (Bernstein, 1975) of the major conventional topic headings in human and physical geography but as a continuum of environments from the dominantly physical to dominantly human. Within the continuum, four key themes were identified where people/environment relationships were fundamental to human life on earth. The headings of the four themes were not static phrases drawn from conventional divisions of the subject, but dynamic, such as 'The use and misuse of natural resources', 'Managing human environments' and 'Issues of global concern'. The descriptors for the six core modules and for the optional modules sustained this note of active environmental themes. Together with an individual study, some nine modules each taking about six weeks of study had to be covered. With this approach to geography came an approach to learning encouraging a process of increased student autonomy through enquiry-based teaching/learning. A route for enquiry was identified in which a double helix of facts and values spiralled around a spine of key questions.

In partnership with ULSEB, assessment arrangements were developed which followed, rather than distorted, the aims and objectives of the curriculum. Neither teachers nor pupils had to outguess the examiners in a poker game of spot the question; nor was the syllabus so broad that large areas purportedly within it were left unvisited by questions for years on end. Examiners did not have to play the joker and set boundary questions which most teachers and students had avoided in their efforts to cope with an overloaded syllabus, nor did they have to signal changes about the meaning of the syllabus by inserting innovative questions on the run, shocking both candidate and teacher out of complacency at the moment of the examination. Essay questions were both short-timed and extended and, along with a techniques application unit, were set and marked in the schools at intervals during the two-year course and then moderated by the board. All students undertook an individual study and one of the two formal terminal papers set by the board was a decision-making exercise which supported the enquiry focus of the curriculum framework. Coursework contributed 35 per cent of the total marks and, with Paper 2 a structured response paper, none of the terminal marks was based on essay questions. Many teachers pressed for a 50/50 split between terminal examination and moderated coursework. However, by 1981 there was growing evidence of a desire among those responsible for examinations policy to protect A level as the 'gold standard' of the whole public examination system by a re-emphasis upon the separation of powers between teaching and assessment, and a heavier weighting towards coursework was no longer acceptable to the curriculum and assessment gatekeepers.

During the 1980s, the syllabuses of some other boards were revised building on the work done by the Geography 16–19 Project. UCLES began to explore modularity, and its 1992 linear syllabus featured the latest thinking about syllabus specification. The two smallest boards, however, offered the most instructive illustration of innovations with their specialised circumstances arising from their locations in Northern Ireland and in Wales. The syllabus of the CCEA was a model of a succinctly specified A level, incorporating a clear and purposeful distinction between aims and objectives, and a specification grid indicating the weightings between the three elements of learning and the four components that make up the content of the examination. The syllabus unit on development was just one of a number of units where the thinking of the syllabus was modern and multidimensional in tone, and that on ethnic diversity was also original and challenging. The CCEA syllabus also followed Geography 16–19 in an emphasis on enquiry, with a decision-making exercise and a personal geographical enquiry.

Policy change

The development of a subject core

For anyone of a rationalist persuasion it would seem self-evident that any subject can be distinguished by its content, and that at whatever level it is studied there should be considerable agreement about what material is appropriate to it. In practice, reality is less tidy than a Cartesian would wish, but from the Callaghan speech in 1976 at Ruskin College onwards there has been a growing determination in the corridors of power to tackle the tortuous and ungodly jungle of educational provision not only in the 5–16 age band, but beyond. It must be acknowledged that this view at that time had some substance. In geography, for example, an inter-board conference was arranged in 1979 for chief examiners and board officers of the then nine boards, which proved only to highlight the differences of content, procedure, marking and, more dauntingly, of outlook between them.

In 1980 the Schools Council suggested reducing the number of syllabuses at A level, and of identifying common cores of content and skills within subjects. Inter-board working parties were convened in 11 subjects and their reports were published in a *Common Cores* booklet (GCE Boards, 1983). The Geography Working Party discussed and discarded the more obvious approach of equating the core with those elements common to all operating syllabuses. A somewhat arcane apologia followed, leading to recommendations which were broad enough to perpetuate a tolerant catholicity and acknowledgements to the core began to decorate syllabuses as italicised headers in new syllabus revisions as they appeared.

By 1992 (see also Butt, p. 180) the matter of subject cores was pressed once again, but this time with greater determination, even though the government was still very heavily involved with the implementation of the National Curriculum for the 5–16 age-group. Draft proposals for a subject core in geography were circulated for comment during 1992/93, and hopes were expressed that student interest would be stimulated by moving decisively 'from the familiar to the refreshingly different and challenging' (SEAC consultation draft). The core was intended to be a framework, and not a syllabus. Only one-third of any particular A level syllabus, and two-thirds of an AS syllabus, need exemplify the core requirements. Figure 10.3 summarises the threefold requirement of a chosen physical environment, a chosen human environment and an integrative theme, as well as requirements for A level students to undertake personal investigative work, although the use of first-hand data is required only for A and not for AS level.

Once the new subject core had been approved (SCAA, 1993) the boards had to work with great speed and exploit

(A) *Principles. To secure:*
  (1) A knowledge of places which provide a balanced sampling of different spatial scales and locations and emphasise the wide range of environments found on earth
  (2) A capability to incorporate new developments in the subject and the use of information technology
  (3) A capability for accommodating temporal and spatial change
  (4) The ability to link particular cases with general models and theories
  (5) Knowledge, understanding and skills through the study of themes and places
  (6) The development of knowledge, skills, ideas and concepts

(B) *Core*. Requires an integrative study of:

I	A theme which emphasises the interaction of people and their environment	*Prime focus* Relevant systems and processes, outcomes, changes through time and consequent issues, responses and strategies
II	A chosen physical environment	Its characteristics, processes—terrestrial, atmospheric, biotic and human—their interaction, consequent spatial outcomes and changes over time
III	A chosen human environment	Its characteristics, processes—economic, social, political, cultural and physical

and additionally the undertaking of:

IV	Personal investigative work	To be based on first-hand* and secondary data *A level only requirement within the core

Figure 10.3   A/AS Subject Core: Geography. *Source:* SCAA, December 1993

the goodwill of many senior examiners to revise their syllabuses to meet the new requirements in time for implementation from September 1995. The NEAB syllabus revised for first examination in 1995 had not even been taken before further revision was required. At a practical level there were three major problems: to produce syllabuses which were co-teachable for the first- and second-year students in schools of small size, to lock together A and AS, and to meet the requirement, originating in the *Code of Practice* for A and A/S, that 'syllabuses will provide for a form of "synoptic assessment" which tests the candidate's understanding of the connections between the different elements of the subject' (SCAA, 1994, p. 9). An additional problem was the growing pressure to produce syllabuses suitable for modular assessment despite their complexity and administrative cost.

Modularity

It may seem strange that modularity had not been adopted rapidly by many A level boards after Geography 16–19 had employed it within the framework of the four themes, with each module representing half a term of study. Its *curricular* advantages for A level were discussed some years ago (Hall, 1986). It placed a corset on the pressures to expand content at the expense of enquiry learning, adding realism when a syllabus was constructed, and allowing both teacher and pupil participation in choice, content and ownership of the course of study.

Some developments did take place. For example, using Geography 16–19 as a guide, the WJEC introduced its new syllabus in 1985–87 with six compulsory board-based core modules, and four further published option modules. A pilot A level was also launched by a consortia of local authorities in the South-West across a range of subjects including geography which became known as the 'Wessex Modular Project', with the assessment arrangements being administered by the AEB. Many modules were designed to introduce work directly related to industry, and to explore route maps linking Wessex modules with BTEC and later GNVQ courses.

The modular developments of Geography 16–19, WJEC, Wessex Modular, and UCLES were matters of syllabus structure only, of developments of curriculum and not about fundamental changes of assessment practice. Geography 16–19 was unusual in having internal examinations marked by teachers and moderated by the board during the two-year course, contributing 35 per cent towards the total assessment. In many syllabuses geography coursework, usually in the form of individual studies, was submitted during the second year. The marks were assembled at the award stage with those of other papers for their contribution to the subject results announced in August.

The SEAC (1992) *Principles* document indicated new ground rules for assessment of modular syllabuses which would distinguish them from linear courses. Modular syllabuses were to be A level courses which could be assessed at different times of the year and the results of modular building blocks reported and banked for up to four years for credit towards an A/AS level award. No module could be worth less than one sixth of the A level syllabus, and, as with linear syllabuses, coursework was limited to 20 per cent of any syllabus. However, at least 30 per cent of the assessment had to be by end-of-course examinations.

Advocates of modular structures for both curriculum and assessment consider them more efficient than linear structured syllabuses in the use of teacher resources, more flexible as instruments in the planning of courses, and more appropriate for student motivation by providing a series of short-term objectives. Modularity is also seen to offer a routeway connecting A level with GNVQ unit-based modular awards. Boards are caught here by market forces in the fear that total entries may drop if a modular route is not offered, and a reluctance if a switch is made to impose quite sweeping syllabus changes on centres who may then move to another board still offering a linear syllabus.

However, modular assessment structures as shaped by the *Principles* have created some practical difficulties in constructing new syllabuses. In order to avoid potential fragmentation of the subject, provision has to be made for synoptic assessment where the subject is examined across a range of content in the syllabus. A maximum of six assessment modules is allowed, of which two modules must be used to meet the core requirements, and form one-third of the total A level syllabus or two-thirds of an AS. If all modules carry equal weighting, then coursework, which is likely to be the personal investigative study, will be limited to one-sixth rather than the permitted 20 per cent maximum. If coursework rises to the full 20 per cent, the five other modules will carry less weight or will have to be reduced to four.

Boards which see modular developments as a response to the 'far reaching changes occurring in Britain's education system today' (UCLES syllabus) suggest this regular and more frequent assessment leads to greater motivation and provides for diagnostic feedback as well as giving new possibilities to test in a wider variety of ways. They see an opportunity for students to design their individual pro-grammes as their experiences and results assist in the ongoing process of their A level course of study. Kathleen Tattersall (1995), Chief Executive of NEAB, believes the inherent flexibility of modular courses will make the A level more accessible and no less demanding than conventional courses and, provided the GCE board is satisfied that a meaningful course is tenable, cross-curricular modules will be validated.

Unfortunately for Wessex, its subject components had been split into 10 modules (10 per cent equal weighting), with 6 modules forming the core. In consequence there was a grinding of gears from Bristol to Truro, and particularly at Chard. With the withdrawal of support for further development work the project was abandoned and with it the pilot attempts to bridge the academic/vocational divide by linking A levels with vocational courses. It is also probable that some recently modified modular syllabuses will have difficulty in linking with GNVQ where the weighting of modules is uneven, thus complicating the comparability of the two routes for accreditation of success for admission to higher education.

AS levels

GCE Advanced Supplementary Examinations were introduced in 1988 in a move to encourage breadth in the curriculum. Like Edmund in Lear, the idea of a half of an A level has reappeared after a long absence in a long play to breathe the words 'the wheel is come full circle; I am here', for quite literally it can be traced back to the pink papers of the Higher School Certificate Subsidiary Level Examination which died in 1950. But policy about broadening the curriculum has been confused and piecemeal, with the government never having come to grips with the arguments in favour of a balanced curriculum across the major areas of experience as is the case in the Swedish Laroplan, or the International Baccalaureate. Every constructive recommendation from the Schools Council (1973) N & F levels to the Higginson Report (1988), proposing once again the abolition of the three-subject A level system, was set aside in the fear that the 'gold standard' of A level would be debased.

It is hardly surprising that AS has been beset with problems since its inception. In particular, the government required that AS should be of an equivalent standard to a full A level and should undertake a contrasting role in a student's programme of study. Early surveys by SEAC (1989) found that a large proportion of candidates were sitting the examination at 17 at the end of their first year after GCSE, and individual students' subject combinations had a strong complementary, rather than contrasting, pattern. Examination boards were vigorously reminded that parity of AS with A level should not be undermined by a relaxation of standards in the grading process.

In geography, AS level entries became no less a disaster than in most other subjects, with boards spending much time and resources developing and examining papers with derisory candidatures. In 1994 a total of only 885 candidates entered the AS examination in geography, with some boards able to count the entry in tens and declining! There is little evidence that admissions to universities have been biased against AS qualifications, even if the reported difficulties might reasonably have led to some unease. It may well prove that the recent cross-curricular modular developments described in the last section will offer a *tabula rasa* and avoid the difficulties AS has experienced in the search for breadth.

AS has, however, added unwelcome constraints upon the development of the new A level syllabuses. It is difficult enough to plan an A level syllabus which meets all the new core requirements without having the complications of co-teachability with AS and transferability of credits across from AS to A level awards. First-hand investigation required at A level is in danger of being reduced in AS to the mimicry of an enquiry using second-hand data, not the best feature of some current (1995) syllabus options. It is also hard to see how the arguments for coherence through synoptic assessment can reach equivalence with A level, when AS has only half the number of component contents to support it. Only two positive comments can be added. OCSEB and UCLES co-operated to produce one common AS. Also the AS syllabuses created by most boards some five years ago incorporated developments in the specification of examinations, thus preparing the ground for improvements in this respect at A level, as new syllabuses were formulated.

Contemporary problems and issues

Comparability of standards

The issue of comparability of standards has long been a matter of concern. It is important to distinguish four aspects to it:

1. Maintaining standards over time in respect of the A level as a single examination system.
2. Maintaining standards one year with the next in the particular subject syllabus of a particular board.
3. Comparability of standards between subjects.
4. Comparability of standards between syllabuses in the same subject offered by one board, and between boards.

Finally, there is the problem of the equivalence of standards reached by candidates within one subject syllabus where the level of demand may vary between optional questions within a question paper or between sections of a paper or between components which are optional.

Norm referencing has long been seen as the core mechanism which protects standards over time, and allows standards to be compared. It allows performance to float, but fixes the proportions of candidates achieving each particular grade on the argument that for any population of candidates, the innate distribution of intelligence remains constant. Its use in the 'broad view' of assessment has already been discussed, but it is also indispensable in addressing the more general aspects of comparability listed above. Figure 10.4 shows the wide variation in the grade percentages gained by candidates in one subject compared with another. A recent study reports that physics is the most difficult subject, followed by chemistry, maths and biology, and that one reason for their higher level of demand for a particular grade level is the selective nature of their candidate entry, which in turn raises the threshold standards for each grade (Fitz-Gibbon and Vincent, 1995). This highlights the inherent difficulty in extending norm-referenced comparisons across subjects, and leads towards attempts to diagnose the nature of the populations entering each subject by testing general abilities such as verbal and spatial reasoning.

Comparability based on norms may be on more secure grounds when looking directly at one subject only. In geography, there has been an upward drift in the proportions of grades awarded, with 12 per cent achieving an A grade in 1994, and nearly 30 per cent A or B. The overall success rate has risen to 80 per cent which is 10 per cent higher than the original norm set for the E/N boundary. Differences in grade percentages between examination boards in the same subject, and between one subject syllabus and another, have not been subject to rigorous examination, but wide variations

	A	B	C	D	E	N	U	Sat exam	% total A level
Art and Design	13.3	31.7	54.4	75.6	90.5	97.5	100	35 350	4.8
Biology	13.0	28.4	44.1	61.3	77.4	88.9	100	47 320	6.5
Chemistry	16.7	34.9	50.9	66.5	79.7	89.6	100	40 772	5.6
English	12.9	30.9	51.1	72.1	87.4	95.0	100	88 739	12.1
French	18.5	35.8	55.2	73.8	87.5	95.1	100	29 637	4.1
History	12.5	29.4	48.6	68.0	82.9	91.6	100	46 096	6.3
Mathematics	24.3	41.3	56.9	70.7	82.0	90.0	100	64 676	8.9
Physics	16.7	32.7	48.9	65.8	80.1	90.4	100	37 349	5.1
Geography	11.8	28.5	46.1	65.2	80.9	90.9	100	46 399	6.3

Figure 10.4 Cumulative percentages of subject results by grade in selected subjects (UK candidates only), June 1993

between boards do exist. For example, NEAB grade patterns in geography have remained close to established norms, while in recent years OCSEB has placed in excess of 20 per cent of their geography candidates in the A band and 25 per cent in the B, raising the top two bands cumulatively to nearly 50 per cent of the total entry.

Criterion-referenced grading might be employed, using general criteria in attempts to relax the grip of norm referencing as an absolute guarantee of unchanging standards. In principle, it reverses the parameters employed in norm referencing by fixing the levels of performance and allowing grade percentiles to float. Where syllabus objectives are explicitly stated, and different papers or sections of a paper perform specific functions and contribute to only some but not all of the objectives, criterion referencing can be of value. But the broad criteria which can be applied to all answers of a particular type, such as free response, are unhelpful in ranking candidates in the narrower circumstances where the instruments of assessment may be stimulus response or data response, or where resources are distributed some days before the actual examination is taken. The cost is a redefinition of the meaning of the 'gold standard', in that the criteria of performance change in response to the changing nature of the subject, or vary from one syllabus to another where different perspectives of the subject are offered. The weightings and prescribed headings of the cross-curricular elements of learning also vary in response to the forces seeking to change their relative values. Thus the quality of excellence becomes contingent upon the dynamics of a social culture rather than of universals inherited from an unevenly distributed biological endowment.

Suggestions that there should be a new A* grade, similar to that now a feature of GCSE, appear to have lapsed, and the old scholarship papers, which used to function as additional information for admissions tutors long after their original purpose as instruments in the competition for means at university had gone, have withered in parallel with the disappearance of the third-year sixth form. Even so, the OCSEB new syllabus retains an optional Special Paper 0 which draws on the whole syllabus, emphasising contemporary geography and its literature, although its intended function remains unclear. Perhaps it reflects the desire of the chief examiner to retain in the school curriculum that essential element of speculative scholarship, a spark drawn from the light of Hellenism to counterbalance the weight of scholasticism in a Hellenistic world.

The new generation of syllabuses

From the autumn of 1995 all students starting A/AS levels took the new syllabuses which, in one way or another, incorporate the subject core. Figure 10.5 shows modular assessment has been adopted by all boards, and only UCLES and OCSEB have retained syllabuses which offer no alternative to end-of-course examinations. Most boards did not radically alter the contents of their existing syllabuses, but many had problems with matching certain requirements of the core and the strict assessment requirements of the *Principles*. UCLES, for example, protected the existing content, components and question styles of its linear syllabus from radical revision, and the core requirements were matched against specific sections of particular papers. A more innovatory approach to content change and extended assessment over time was taken by the new UCLES modular syllabus, which is distinctive in that it 'paints the study of geography onto a global canvas, where global systems, links, and patterns can be appreciated'.

The NEAB syllabus also takes an innovative approach to content. The core concept of change, the emphasis on short and long time-scales, interactions, core skills and enquiry, underpin the treatment of the physical environment (short

Syllabus	Assessment availability	Length of investigative study	Weight (%)	Assessed by	'Written-up' timed paper option to individual study
AEB 0626	Modular and linear	4000	20	School	Yes
NEAB Geography	Modular and linear	3000	16·7	School	Yes
OCSEB 9630	Modular and linear	4000	20	Board	No
UCLES 9050	Linear	4000	20	Board	Yes+report
UCLES 9518	Modular and linear	4000	16·7	Choice	No
ULEAC 9201	Modular and linear	2500/3500	20	School	No
ULEAC 9211**	Modular and linear	4000	20	School	No
UODLE 9945	Modular and linear	2500	18	Board	No
WJEC Geography	Modular and linear	3000	16·7	School	No

*First examination June 1996; modular syllabus in preparation
**ULEAC Syllabus B—Geography 16–19 project

Figure 10.5 Details of A level syllabus approved for 1997. *Source:* SCAA, 1996

and long term), the treatment of interactions (the conflict over resources, hazards) and the human environment (the United Kingdom and the wider world). The major adjustment to this syllabus has been its reshaping into modular form, with six modules of equal weighting, two related to personal enquiry and decision-making, two on syllabus core content, and two on optional studies.

The synoptic assessment requirement, referred to above, caused greatest difficulty and confusion. It was not clear whether synoptic assessment referred to the subject itself, or to the subject syllabus being submitted for approval, and different syllabuses have met that requirement in fundamentally different ways. For example, with the UCLES modular syllabus, it is a free-standing module specifically designed for this purpose entitled 'Geography in action within a European framework'. In contrast, the decision-making paper of the new UODLE syllabus draws upon its core modules 1 and 2 of the syllabus to meet the call for synoptic assessment.

Fieldwork remains a difficulty, with numerous in-built tensions. There is a conflict between the core requirement that geography syllabuses must include personal investigative work, and the absence in the A and AS principles of compulsion in the assessment arrangements. Some boards have therefore introduced a timed examination paper as an optional alternative to the submission of a personal folder. This route, which attempts to simulate personal enquiry through the mimicry of a timed written examination paper, must be highly questionable. Past experience has produced unstable results, even allowing for any redefinition of fieldwork brought about by modern technology. The concession that AS need only offer secondary data has added to confusion, with the arguments over what actually is meant by first-hand data collected in the field, and the point at which data become secondary.

For Geography 16–19 the restriction of coursework to 20 per cent has been a particular blow to the curriculum framework. The retreat from the concept of enhanced responsibility for teachers in the assessment process has put at risk over 20 years of development work. It ignores the research done by the WJEC on the GCSE Avery Hill Project, showing the strengths of such assessment practice where moderation procedures are well tuned and management effective. It can only be hoped that circumstances will bring about a change of policy before all the personnel and impressive expertise of the team of moderators built up by ULSEB is lost.

The future agenda

The insemination of a core requirement at A level signals a turning point in the relationship between the regulating authority and the executive institutions charged with the origination and administration of particular syllabuses. The School Curriculum and Assessment Authority (SCAA) is now the *pole phare*, responsible for the initiation of policies of change and the monitoring of procedures to make change effective. The eight boards themselves, as *poles d'ancres*, have co-operated with SCAA to produce an agreed *Code of Practice* (March 1994) that is intended to promote consistency and quality in the examining process and to help ensure that grading standards are consistent in each subject across the different boards, and between different syllabuses in the same subject in one board. Common approaches to question setting, revising and the drafting of mark schemes are outlined, and particular attention is given to the quality of language in the marking of scripts. All syllabuses will track a statement of objectives through to the questions set and the mark scheme by means of a specification grid to audit the consistency, accuracy and fairness of the examination. It has also been agreed to harmonise the procedures at the award, grading the judgemental boundaries component by component. Mark schemes are to be published. Language is another priority, with guidance recently issued on this (SCAA, 1995). Clarity of expression, structure and presentation of ideas, grammar, punctuation and spelling are listed as the elements of quality. Statements about language quality must be included, both in syllabus objectives and in the rubric of question papers requiring continuous prose answers.

For over a decade scrutiny teams have reported on their monitoring of examination procedures on a subject basis, but up to 10 years could elapse between visits. In 1993 plans to undertake simultaneous scrutinies in one subject across a number of examination boards were implemented, and geography was scrutinised on this basis for the first time in 1994. Undoubtedly the core is intended to facilitate comparability of standards between boards, and it is part of government strategy to use subject cores as a common basis upon which comparisons can be made and standards confirmed and regulated.

Yet the last section has shown that the new syllabuses are hybrid. The systematiser will contrast the treatment of the physical core by OCSEB, which requires a compulsory short answer response across all the core, with some other boards where the basic processes of the lithosphere, atmosphere, hydrosphere and biosphere are specified in the syllabus, but then treated as options in the examination papers. There is little agreement about the content descriptors in the human geography core, being population and settlement in one syllabus, population and food supply in another, and population–resource relationships and urban processes in a third. Some syllabuses, quite properly when considered against current trends in geography (Walford and Haggett,1995), rotate these frames for description into more dynamic and contemporary conceptualisations, as with CCEA and NEAB. Indeed, the writer drafted an innovative module 'Movement and the environment' in response to the challenge of the core document to 'introduce new and more demanding approaches' only to have it gently set aside as it became clear that only five modules could be incorporated without transgressing the *Principles*, and that teachers would not have the resources or the time to manage it effectively. It rests as one more testament to the weight of the ballast of inertia, unless it can be slipped in surreptitiously as an option module. One wonders how we can prevent our views of geography being stuck in a time warp of rural England with its corn-dollies, home-made jam, and cottage cheese, when the real world is alive with the music of Pentium-serviced personal computers and Internet conversations.

A regrettable step in assessment practice has been the conflation of the elements of learning into two descriptors, (1) knowledge and understanding, and (2) skills, with concepts, attitudes and values subsumed under knowledge and understanding. This simplification is unfortunate when the present need is for chief examiners and teachers to work to a specification grid, which is sufficiently broad to link powerfully the assessment objectives with the questions set in the examination papers, and the weightings for the reward of marks in the mark scheme. The recently revised syllabuses from UCLES and NEAB had provided weightings for three major headings to separate out values and attitudes from knowledge and understanding with positive effects on the setting of questions and mark schemes.

Attitudes and values have suffered subordination because they are presently considered by powerful interests to undermine rigour and so debase standards. This view turns its back on the constructive work done by the HMI in the Red Books between 1974 and 1983. At that time deliberations drawn from curriculum theorists from Tyler (1950) and Taba (1962) onwards were translated from matters for discussion to matters for action by curriculum developers including HMI (DES, 1983, 1985) so that these elements of learning became central to the tasks of syllabus development. Unfortunately, it is also possible for critics to dismiss such approaches as a numbers game played in the abstract offices of theorisers. It is important to be clear that their purpose is to enhance the quality of the the assessment process, and not to interpret the specification grid as a positivist approach to knowledge. A scan of the rationales, the aims and the objectives offered in the newly approved syllabuses shows how much remains to be done in applying principles of curriculum and assessment planning, and in creating a dialogue between chief examiners, in order to reach some common basis about the meaning and function of these terms. Its importance is emphasised by the wholesale transfer of aims and objectives from the syllabus of one board to that of another to meet the core revision deadlines.

Overall, the application of knowledge of curriculum theory in the subject-driven world of the examination boards remains weak. The major workforce is casual, founded on the irregular goodwill of professionals whose main duties lie elsewhere in subject teaching and research outside the increasingly specialised domain of assessment matters. University lecturers from subject departments can no longer be recruited to fill key positions in the examining process, and the long-standing relationship between school and university departments in matters of assessment is being blown away by the winds of changing career patterns and priorities. Boards will find it increasingly difficult to recruit personnel of sufficient experience to undertake examining tasks which have become more challenging and time consuming, and who yet require training if they are to respond effectively to the *Code of Practice*. Key personnel, used to taking a broader view of assessment procedures, and with the instinctive professionalism of the amateur of long experience, will step down, seeing the control passing from subject expert to the bureaucrat, and the passing away of that essential flair in setting, marking and awarding examinations which has been held to transmit what is worth knowing from generation to generation. For, since its inception, the whole system has been serviced by a sense of *noblesse oblige*, where an honorarium is given rather than a salary paid. But as the roles of chief examiner, moderator and reviser are becoming a matter of technical expertise as much as of a love of subject, proper remuneration can no longer be dismissed as an ungentlemanly concern with money. As the personal secretary to Charles I said as the Royal Standard was raised at Oxford, 'there is more to it than bidding it be done'.

For nearly half a century, A level has retained a unique role as the determinant of the curriculum of young people in schools and further education colleges who have chosen the routeway leading towards higher education institutions. It has been the major selection mechanism for competitive entry to universities. But government targets for greatly increased numbers of university students by the year 2000 have already been exceeded. Today there is force in Sir Geoffrey Holland's (1995) view that in terms of tomorrow 'the over-narrow A-level is the one main area . . . that cries out for reform'. Far from being the 'gold standard', as ministers tend to regard them, they are in fact an altar on which have been sacrificed the enthusiasms, the hopes and the capabilities of about half our young people in the past. A levels are long past their sell-by date. What we need is something now which links together the academic and the vocational. The Dearing Review of 16–19 qualifications has, during 1995, embarked on that task.

Yet comparison of an A level module which the writer has written on leisure and tourism with its GNVQ equivalent is a disturbing example of the difference in outlook on what is worth learning, the contrast between a narrowly conceived entrepreneurial and a broad globally framed ecological perspective. In the 16–19 curriculum we need to replace the image of the guinea with that of the ECU. In geography at least some syllabuses are moving in this direction in both subject-specific and cross-curricular terms. The urgent national priority is to melt down our major subject frames and recast them in a broader curriculum code which integrates learning across the intellectual breadth of the contemporary culture. If we could catch the spirit of the International Baccalaureate, we would escape from the narrow provincialism and outlook which the Crowther Report placed upon us some 40 years ago, and which has conditioned subsequent thinking for far too long.

References

Bernstein, B. (1975) *Class, Codes and Control*, 3. Routledge & Kegan Paul, London.

Benjamin, H. (1939) The saber-toothed curriculum, reprinted in R. Hooper (ed.) (1971) *The Curriculum: Context, Design and Development*. Oliver & Boyd, Edinburgh.

DES (Department of Education and Science) (1983) *Curriculum 11–16: Towards a Statement of Entitlement* (The Red Book). HMSO, London.

DES (1984) *AS Levels*. DES, London.

DES (1985) *The Curriculum from 5–16. Curriculum Matters 2*. HMSO, London.

Fitz-Gibbon, C.T. and Vincent, L. (1995) *Candidates' Performance in Public Examinations in Mathematics and Science*. SCAA, London.

GCE Boards (1983) *Common Cores at Advanced Level*. The GCE Boards, Oxford.

Hall, D.B. (1986) Advanced Level examinations. In D. Boardman (ed.) *Handbook for Geography Teachers*, Geographical Association Sheffield, pp. 257–268.

Hall, D.B. (1990) 'The National Curriculum and the two cultures: towards a humanistic perspective. *Geography*, 75(4), 313–324.

Hall, D.B. (1991) Charney revisited: twenty-five years of geographical education< In R. Walford (ed.) *Viewpoints on Geography Teaching*, Longman, Harlow, pp. 10–29.

Hall, D.B. (1994) Postgraduate student understanding of the specifics and professional dimensions of the National Curriculum in England and Wales. In H. Haubrich (ed.) *Europe and the World in Geography Education*. IGU Commission, Nuremberg, pp. 297–324.

Higginson Report (1988) *Advancing A Levels*. HMSO, London.

Holland, G. (1995) *The knowledge programme: 2020 vision*. BBC2 21.3.95 (reported in *The Guardian*, 10.3.95).

Naish, M., Rawling, E. and Hart, C. (1987) *Geography 16–19; the Contribution of a Curriculum Project to 16–19 Education*. Longman, Harlow.

OFSTED (Office for Standards in Education) (1993) *GCE Advanced Supplementary and Advanced Level Examinations—Quality and*

Standards. OFSTED, London.

Reid, W.A. (1978) *Thinking about the Curriculum*. Routledge & Kegan Paul, London.

Robbins Report (1963) *Report on Higher Education*. HMSO, London.

SCAA (School Curriculum and Assessment Authority) (1993) *GCE A and AS Subject Core for Geography*. SCAA, London.

SCAA (1994) *Code of Practice for GCE A and AS Examinations*. SCAA, London.

SCAA (1995) *Assessing Quality of Language*. SCAA, London.

Schools Council (1973) *16–19 Growth and Response: 2. Examination Structures*. Working Paper no. 46, Evans Methuen, London.

SEAC (School Examinations and Assessment Council), (1989) 'A-level consultation and AS survey'. *SEAC Recorder no. 3*. Autumn, 2–5.

SEAC (1990) *Examinations Post-16: Developments for the 1990s*. SEAC, London.

SEAC (1992) *Principles for GCE Advanced and Advanced Supplementary Examinations*. SEAC, London.

Taba, H. (1962) *Curriculum Development*. Harcourt Brace, New York.

Tattersall, K. (1995) Reforming the Gold Standard. *Guardian Education*, 21.3.95.

Tyler, R.W. (1950) *Basic Principles of Curriculum Development*. University of Chicago Press, Chicago.

Walford, R. and Haggett, P. (1995) Geography and geographical education. *Geography*, 80(1), 3–13.

Developments in geography 14–19: a changing system

11

Graham Butt

Introduction

The 1990s will be remembered as a time of major change in the organisation, assessment and accreditation of education for the 14–19 age range. These developments affect geography education in numerous ways, combining to create significantly different educational choices for children from those which existed previously. Decisions made at 14 concerning the academic or vocational pathways taken by each child naturally affect subsequent educational choices and experiences up to, and beyond, the age of 19. The importance of offering courses containing viable geographical education components to children at times when these decisions are made is obvious, especially as the Dearing (1993b) Final Report on National Curriculum and Assessment encourages the increased provision of vocational pathways from 14 to 19. A further Dearing Report in July 1995 is focused specifically on 16–19 qualifications and may result in further changes in emphasis after 1996 (Dearing, 1995).

Although there have been many proposals for a more unified curriculum and assessment system for this age group, three educational pathways currently exist post-16—the 'academic' route traditionally occupied by GCE Advanced (A) levels, the 'vocational' route linked directly to education in the workplace (such as National Vocational

Geography into the Twenty-first Century, Edited by E.M. Rawling and R.A. Daugherty,
© 1996 John Wiley & Sons Ltd.

Qualifications (NVQs)), and the 'mixed' route of General National Vocational Qualifications (GNVQs). The latter have been created to help students keep their educational options open for longer, for GNVQs can be geared either to employment, training or further and higher education.

The 'academic' route from 14 to 19

The Dearing Final Report, GCSEs and geography

The necessity for geography departments in schools to consider seriously the vocational education debate has been heralded by the Final Report of Dearing (1993b), in which geography and history became optional subjects after the age of 14 (see Dearing, 1993b, paragraph 5.25). The reasoning that despite both subjects being 'valuable' they should not be given legal priority over 'creative arts, a second foreign language, home economics, the classics, religious studies, business studies or economics' is perhaps unclear. Changing the compulsory status of history and geography at the end of Key Stage 3 had little to do with the nature of these disciplines—it was more closely associated with the need to create additional curricular space within Key Stage 4 for other academic and vocational options. As a result, both subjects were returned to their original status of General Certificate of Secondary Education (GCSE) options at a time when most schools were also under 'external pressures to develop vocational courses' (Bennetts, 1994, p. 63).

Great concern was expressed in the early 1990s that the introduction of new GCSE syllabuses in geography (to coincide with the expected start of National Curriculum Key Stage 4 in September 1994) would create a confusing rush for pupils, teachers and examination boards. Geography GCSEs had largely been a success following their introduction in the mid-1980s, and many teachers were nervous of ill-considered and overhasty change. It was also unclear how GCSEs could be devised to assess the plethora of statements of attainment contained in the existing Geography Order; this problem was unceremoniously passed to the examinations boards who inevitably found themselves 'investing considerable time and money in the development of GCSE syllabuses' (Orrell and Wilson, 1994, p. 90).

However, in October 1993 the then Secretary of State for

Education, John Patten, announced a two-year postponement of the launch of new geography and history GCSE syllabuses, ostensibly to relieve pressure on teachers. Orrell and Wilson (1994, p. 91) suggest that this postponement was also partly due to the existing GCSE syllabuses, from which boards were devising their new courses, 'not [being] the ideal sequel to KS3 teaching programmes'. With a more realistic time-span for the introduction of the GCSEs the fears of poor implementation, teaching and examination performances—possibly leading to reductions in future candidate numbers for A and AS level geography—were largely dispelled.

Most boards were left though with almost completely reworked GCSE geography syllabuses tied to defunct statements of attainment for a final key stage in geography that no longer existed. The task for the examination boards was therefore refocused into providing GCSE syllabuses for 1996, including provision for the 'starred' A grade, based on revised GCSE criteria for geography published in January 1995.

The Dearing Final Report (Dearing, 1993b) contained particularly strong recommendations for vocational education options to be more prominent from the age of 14. With statutory subject requirements in the National Curriculum only occupying 60 per cent of curriculum time at Key Stage 4 the 'gap' created for vocational courses and optional GCSEs to compete for the remaining 40 per cent of time became considerable. Dearing supported the notion of delaying undue specialisation and providing vocational pathways from within this key stage, thus creating a clear focus upon a 14–19 orientation of the curriculum.

The 'three pathways' (A level, NVQs and GNVQs) in post-16 education have therefore now become extended 'downwards' to incorporate the 14–16 age range. Dearing (1993) was keen for the School Curriculum and Assessment Authority (SCAA) to discuss with the National Council for Vocational Qualifications (NCVQ) the possibilities of GCSE and GNVQ accreditation combining to strengthen the academic and vocational balance of education within, and beyond, Key Stage 4. Bennetts (1994, p. 60) expected this aspect of Dearing's proposals to be 'a setback for geographical education', while Rawling (1994, p. 98) highlighted the important issue of geography continuing to maintain a prominent place in the curriculum. She noted that 'the core subjects start with a built-in advantage, because they receive less slimming and a greater amount of curriculum time at all key stages', which combined with more rigorous testing arrangements gives the impression that geography is a subject that is 'less important and so less deserving of resources, staff and curriculum time . . . the first candidate(s) for downgrading in whole school priorities'.

Whether children of all abilities will find the vocational alternatives offered at 14 more attractive than continuing to study 'straight' geography is, at present, a matter of speculation. Geography's popularity within options at 14 was witnessed by its high entry numbers at GCSE in the early 1990s. However, children's educational choices into the next century are potentially much wider given the increasing breadth of vocational opportunities.

Geography departments in many schools will certainly need to continue to fight for curricular space at Key Stage 4, and will require exciting and relevant courses at Key Stage 3 to make children want to opt for geography. The reasons why children should carry on studying geography must therefore be made clear to pupils, parents, employers and other teaching staff.

Developments 16–19

At the beginning of the 1990s developments concerning the education of 16–19-year-olds were occurring as something of a 'quiet revolution'. Teachers largely focused their attention on the demands of the National Curriculum, the Dearing Reports (Dearing, 1993, 1994), and the pronouncements of SCAA (1994). Meanwhile important changes suggested, and indeed implemented, with respect to post-16 education, failed to create the impression that they might have done in less interesting times. Not least among these were 'changes in geography syllabuses, methods of assessment, the academic and vocational balance of courses and the question of modularisation' (Butt, 1993 p. 36).

Since well before the Higginson Committee reported in 1988, the future of the main academic qualification post-16—the A level—had been under review. The government was keen to advance the debate along certain selected lines, as witnessed by its White Papers *Education and Training for the 21st Century* (1991) and *Competitiveness* (1994). These made clear cases for the development and expansion of vocational courses, primarily in the 16–19 age range, following the creation of the NCVQ in 1986. The NCVQ was established to bring an end to the confusion created by over 300 bodies awarding vocational qualifications, and to instigate common arrangements for the recognition and accreditation of courses. As a direct result the NVQs and GNVQs were developed and administered through the existing awarding bodies of the Business and Technology Education Council (BTEC), City and Guilds (C&G) and RSA Examinations Board.

Some see the 1994 White Paper, *Competitiveness*, with its introduction of a system of vouchers or credits for 16-year-olds to exchange for academic or vocational education, as putting the long-term future of A levels in doubt (Pyke et al., 1994). These initiatives, based on increased modularisation and the promotion of core skills for both A levels and GNVQs, require the creation of simple methods of comparing academic and vocational achievements. Blackburne and Nash (1994) also noted the approval by SCAA of a new vocational AS level at this stage which inevitably raised questions about the government's favoured vocational education route post-16. They argued that if GNVQs have 'parity of esteem' with A levels, why should examination boards be encouraged to venture into vocational AS levels?

Interestingly, despite its newly found concern for all aspects of vocational education, the government still publicly referred to A levels as the educational 'gold standard' in the early 1990s. By 1995, it had become apparent that the whole framework of 16–19 qualifications needed reviewing, specifically to consider ways in which it could be strengthened, consolidated and improved. Sir Ron Dearing was once again commissioned by the incumbent Secretary of State for Education to carry out this review, following his successful recommendations on the National Curriculum. Significantly, a major element of his brief was 'to maintain the rigour of A levels'.

A and AS levels

The A level geography examination is broadly considered to be a success by the government, SCAA and the A level boards. However, since the 1960s A levels have been under increasing pressure to change, so as to create room for the introduction of qualifications applicable to the entire ability range at 16–19. Criticisms that too many candidates who take A level examinations fail them, that progression from GCSE in some subjects is difficult, and that sizeable differences exist between boards in their pass rates and grading proportions have all been raised. Indeed SCAA attempted to address some of these issues in 1994 and 1995 through an A and AS level comparability exercise. There have also been concerns that A levels do not prepare students adequately for many degree courses.

A plethora of possible solutions to the A level 'problem' has been advanced since the mid-1960s. These range from the introduction of 'N' and 'F' levels in the 1970s, to the creation of modular and bridging courses, to the adoption of the International Baccalaureate or its vocational equivalent the *baccalauréat professionel*.

In 1994 some 46 347 candidates took A level geography, making it the sixth most popular A level subject. Overall A level candidate numbers had increased by over 13 per cent in the first half of the 1990s, while during the same period the number of candidates taking A level geography had risen impressively by over a quarter (see Appendix 3). This rate of increase perhaps reflects the popularity and success of GCSE geography syllabuses introduced in the 1980s.

By comparison the AS level, which was introduced in 1987, has not been so successful. It was designed to be of a similar standard to the A level, but was supposed to require only half the teaching and studying time. Unfortunately it proved to be an unpopular examination from the start. AS level geography recorded insignificant numbers of candidate entries compared to the A level (less than 1000 each year in the early 1990s) and many teachers are still concerned about promoting a qualification whose acceptability to higher education institutions (HEIs) is not altogether clear. This is a reasonable fear, for although admissions tutors welcomed AS levels in 1987 few institutions have comprehensively followed words with action. Universities still tend to make their offers primarily on the basis of A level passes.

For a time it seemed the AS level might become 'an engine for genuine curriculum reform' (Swatridge, 1994), but this engine stalled due to small sixth forms not being able to timetable or resource the courses. Independent schools have also tended to ignore them, and many parents, employers and students currently have little faith in their worth.

In March 1992 SEAC produced a leaflet outlining its

future principles for A and AS level examinations (SEAC, 1992). It informed the examination boards of the need to phase in new A and AS level syllabuses a minimum of two years after the implementation of Key Stage 4/GCSE requirements for each subject. Although the government was championing the cause of vocational education, SEAC was not addressing the question of how to bridge the academic/vocational divide, or how to create 'parity of esteem' between such courses. To a certain extent SCAA has aimed to rectify this problem, at least with regard to the AS level, with the publication of guidance materials called *Using the AS* (1995). These sought to provide schools, employers and HEIs with information on the role and functions of AS levels, particularly in combinations with other courses and qualifications.

In the case of geography, the A and AS level principles (SEAC, 1992) heralded the introduction of new A levels in September 1995, for first examination in June 1997. The guidelines directed examination boards to produce courses that were predominantly externally assessed through terminal examinations, with moderated coursework elements reduced to 20 per cent. They also stated that a smoother transfer between GCSE and A level courses was required to reduce the perceived academic gap that existed in some subjects.

The restrictions on coursework, in line with the government's decisions in the early 1990s to reduce the weighting of coursework in GCSE courses, was a blow for those boards who had traditionally championed its inclusion at A level. Paradoxically, given the stated parity between academic and vocational courses, the government seemed content for coursework to have a major role only in the assessment of the latter. Many boards effectively attempted to 'sidestep' this problem by favouring the option of producing modular courses, which avoided some of the harsher restrictions on coursework.

The implicit reduction of the status of coursework also sent the unwelcome message to curriculum developers that innovative courses at A and AS levels might now be harder to establish. The limited advantage of such a reduction was seen by some as forcing a reassessment of the role of coursework within examinations. Geographical skills, for example, which could not be assessed easily by written examinations, would now have to be assigned to coursework if they were to be guaranteed coverage. The reduced mark range created difficulties though, for if the percentage mark for fieldwork (say) is reduced below 20 per cent it may be perceived by many to have lost its significance.

Subject Cores

In 1983 the majority of GCE boards combined to publish a *Framework for A level Geography*, which outlined a 'Core' of geography that might be included in their syllabuses (GCE Boards, 1983). In other subjects, the Cores which were developed were heavily content driven and thus the whole exercise was suspended until 1992, when it was relaunched to include geography, chemistry, English and maths. This time there was a greater measure of success, although interestingly the geography Core of 1993 (SCAA) was more restrictive and included a greater element of prescribed content than that of 1983!

The idea behind creating such Cores was to provide a minimum entitlement in each of the subjects, rather than a prescribed content, to help ensure common standards for all A levels. A further consideration was the aim to develop flexibility so that examination boards could still design innovative syllabuses if they so wished.

In each A level subject, a Core of subject knowledge, skills, techniques, abilities and understanding was therefore identified. This standardisation found favour with educationists and examination boards, although achieving the balance of Core and optional content for geography proved to be problematic. Devising a range of means of assessment capable of evaluating the broad abilities and competences of A and AS level geography students was also taxing.

SEAC consulted widely on its suggested Subject Core for Geography in 1992. The Core represented about one-third of an A level (or two-thirds of an AS level) and was generally considered to be well structured, flexible, imaginative and a suitable basis on which exam boards could base their syllabus designs. Some criticisms focused upon proportions of fieldwork proposed, and the lack of emphasis on qualitative work, but these were fairly minor. Most teachers were happy that the Core did not over-represent statutory geographical content, or divide A/AS level syllabuses into an unwieldy 'core' and 'option' split.

In December 1993 SCAA produced its finalised A/AS

Subject Core for Geography which was to be part of all new syllabuses from September 1995. Broadly the Core requires all candidates to develop geographical skills and to gain an in-depth knowledge and understanding of a people/environment theme at different spatial scales. A chosen physical and human environment, and (at A level) personal investigative work based on both primary and secondary data collection are also required. (AS level candidates were not expected to carry out first-hand data collection as part of the Core, but could do so within the non-Core part of the syllabus.) Lack of mention of IT unfortunately put the geography Core out of line with National Curriculum Geography and the geography GCSE where it is clearly stated.

Modules

One of the major issues in the development of the 16–19 system is the extent to which courses should be designed and assessed on a modular basis. Typically, vocational courses are modular in structure while A and AS courses are not. SEAC reconsidered the proposed modularisation of certain A and AS levels in April 1993, asking examination boards and professional associations to comment on its plans. Ground rules were established for a minimum value of 15 per cent to be awarded for end-of-module exams, with at least 30 per cent for terminal examinations. This created modules with enough marks to be credible as 'free-standing' assessments which were potentially transferable to GNVQ level 3. Definitive statements on the future of modularisation are awaited from SCAA, particularly on methods and procedures of credit accumulation and transfer, as well as the structure and timing of assessments.

Not all subjects lend themselves easily to modularisation, and geography could find itself lacking a coherent whole and clear progression if it becomes too fragmented in a carelessly created modular system. Relationships between different elements of a course, and within the whole subject discipline, can become fractured by modularisation. There are real dangers that an ethos of 'credit building' can overtake the educational need of achieving a holistic understanding and competence within a subject discipline— with the increasing trend towards modularisation this may become a major concern. Geographers have always wished to test candidates' *accumulated* understanding of the subject as a whole and would therefore resist undue fragmentation.

Practical issues such as the cost of producing numerous 'option' modules are also important, for examination boards may not have the time, money or resources to do this. However, if the collection of option modules is not broad enough, the Core becomes too dominant and flexibility is lost. A further problem concerns the variation in standards of candidates' work throughout a two-year course. Achievements within modules taken at the start of a course will vary significantly from those taken later when students' ideas and conceptual understanding have matured. Similarly, the performance of full-time, part-time and mature students may vary.

One thing which does appear clear is that modularisation will only fulfil its fullest potential if there is a strong link between modules, credits, A and AS levels and GNVQs. If modularisation does not succeed, then the academic–vocational link probably will not either. Even if modular courses are broadly successful this will still not guarantee an coherent post-16 framework. Any realistic system that can allow interchange between academic and vocational courses post-16 should be supported, for there are dangers that if we 'marginalise geography as a narrowly "academic" subject' (Butt and Lambert, 1993, p. 181) it may lose its foothold completely.

The rapid pace of change across the 14–19 assessment scene showed few signs of slackening through the 1990s, due to the introduction of new A level syllabuses incorporating the Subject Core, the 'reintroduction' of the geography National Curriculum and the launch of new geography GCSEs. Once complete this will hopefully create a greater continuity of geographical skills, knowledge and understanding across the entire 14–19 age range.

The 'vocational' routes from 14 to 19

The GNVQ 'revolution' and its possible effects on geography 14–19

The epilogue of Chorley and Haggett's influential publication *Frontiers in Geographical Teaching* (1965, p. 375) contains the following prophetic passage:

> if geography trims its sails to the vicissitudes of every profitable wind of social and educational demand that blows it is likely to lose any sense of distinctive intellectual purpose, will fail to attract its most necessary growth ingredient (the research student) and is likely to be eventually replaced by or amalgamated with other subjects which serve the purposes of society as well as possessing some intellectual identity.

As one considers the advance of vocational education the possible effects on geography education within the 14–19 age range appear potentially huge. Is there a real danger that if geography becomes too closely associated with vocationalism it will lose its 'intellectual identity'? Or would a failure to realise the importance of moving with the times signal its demise through lack of popularity in the educational market?

For many years geographers in schools and universities have contributed to the development of academic, vocational and pre-vocational courses for the 14–19 age range. In the 1980s geographers were involved in initiatives such as the 16–19 Geography Project, the TVEI-related Geography Programme (TRIST), and contributions to economic understanding through GSIP (Geography Schools Industry Project). There is also an impressive history of collaboration with TECs, involvement with Industry Years, inputs to Certificate of Pre-vocational Education (CPVE) courses and the creation of link schemes and educational partnerships with local industries and businesses. This highlights the ways in which geographers' contributions to vocational education have been developed without weakening the academic rigour and respectability of the subject.

The point is an important one, for the debate has now moved beyond whether geographers should simply input, say, map-reading skills or locational knowledge to vocational courses. There is a growing awareness of a much broader range of contributions that can be made, consistent with the range of geographical skills, knowledge and understanding that practising geographers have access to.

The General National Vocational Qualification (GNVQ)

GNVQs were announced in the White Paper *Education and Training for the 21st Century* (1991) as a joint replacement,

---

Foundation GNVQ
Nine units—3 mandatory, 3 optional, 3 core at level 1
Suitable for post-16 students as a one-year course towards Intermediate stage.
  Also suitable to some 14–16
Comparable to 4 GCSEs at D to G grades

Intermediate GNVQ
Nine units—4 mandatory, 2 optional, 3 core at a minimum of level 2
Suitable for post-16 students as a one-year course, although some students might
  have time for additional units or GCSEs. Also suitable to some 14–16
Comparable to 4 GCSEs at C grade or above

Advanced GNVQ
Fifteen units—8 mandatory, 4 optional, 3 core at a minimum of level 3
Comparable to A and AS levels. Suitable for post-16 students as a two-year course
Other units and NVQ units may be added to broaden the range. May be added to one
  A level
Advanced GNVQ 'pass'—two A levels at grade D or E
Advanced GNVQ 'merit'—two A levels at grade C
Advanced GNVQ 'distinction'—two A levels at grade A or B

NB Core skills include: communication, application of number and IT

Figure 11.1   The structure of GNVQ courses

with NVQs, for all existing vocational qualifications, and to provide a third clear qualification pathway for post-16 education in the 1990s. As the then Secretary of State for Education, John Patten, stated in a speech on 5 April 1993, 'GNVQs will play a central role in our strategy for 16–19-year-olds. . . . In the longer term I would like to see GNVQs developing into a mainstream qualification for young people, catering for as much as half the age group.' He also stressed his belief that GNVQs were not a soft option and that they would 'stand alongside academic qualifications on their own merits'. However, the GCE exam boards were quick to point out that GNVQs were not required to meet the same codes of practice as A levels, despite their supposed 'parity of esteem'. The question of securing such parity is crucial and problematic, for academic excellence and vocational achievement have traditionally been described and rewarded in very different ways.

GNVQs were first piloted in September 1992 at Intermediate and Advanced levels (see Figure 11.1). The Advanced GNVQ was designed to provide 'a genuine alternative to GCE A level qualifications for the increasing number of students staying on in full-time education beyond the age of 16' (NCVQ, 1993, p. 3). Dearing (1993b, p. 48) also encouraged

SCAA and NCVQ to develop 'a high quality GNVQ option (that) can be introduced into schools as an established and well-respected part of the 14–16 curriculum' which they attempted through a joint task group established in 1994.

The Intermediate and Foundation GNVQs occupied the same timetable space as four GCSEs, therefore pupils would not be able to gain a full GNVQ course qualification by the age of 16. Within Key Stage 4 they would either have to gain 'credits' towards completing a GNVQ post-16, or take a range of GNVQ units towards a 'Part 1' certificate, or take GCSEs that transfer credit towards achieving a full GNVQ. However, problems naturally arose because of the very different structure and character of academic and vocational courses.

The Part 1 GNVQ certificate (equivalent to two GCSEs, and utilising 20 per cent of the 40 per cent 'spare' time created for the school curriculum post-Dearing) was strongly supported by the government, but ran into immediate problems. It was supposed to give pupils and schools a variety of course choices and to be the cornerstone of 14–16 vocationalism; however, the pilot planned for September 1995 initially faced delays due to disagreements between the SCAA and the NCVQ over assessment methods and standards (Nash, 1994). The SCAA was concerned that the assessment of Part 1 courses would not be rigorous enough to stand alongside GCSEs. However, the disagreements were settled, the 1995 pilot was launched and indeed ministers agreed to an expansion in numbers of unit titles and participating schools for 1996.

The NCVQ initially stressed that GNVQs should not be offered solely to children with low academic achievement within the National Curriculum. Pupils on vocational courses have to assume the responsibility for many aspects of their learning and must plan, organise, administer and research as part of their GNVQ courses. They also need to compile portfolios of work and understand and participate in assessment systems that are by no means straightforward. As Marvell (1994, p. 176) pointed out, this responsibility may be 'too great for the generally less able'.

No single 'geography' course exists within the vocational framework, but geographical elements are contained within courses such as leisure and tourism, manufacturing, built environment, information technology, and land-based industries (see Figure 11.2). By 1994 some 94 000 students had

Intermediate and Advanced	1992–93	1993–94	1994–95	1995–96	1996–97
Art & Design	Pilot	Publication		Revised pubn	
Business	Pilot	Publication		Revised pubn	
Health & Social Care	Pilot	Publication		Revised pubn	
Leisure & Tourism	Pilot	Publication		Revised pubn	
Manufacturing	Pilot	Publication		Revised pubn	
Construction & the Built Environment		Pilot	Publication		Revised pubn
Hospitality & Catering		Pilot	Publication		Revised pubn
Science		Pilot	Publication		Revised pubn
Engineering			Pilot	Publication	
Information Technology			Pilot	Publication	
Management Studies (Advanced)			Pilot	Pilot	Publication
Media: Communication & Production			Pilot	Pilot	Publication
Retail & Distributive Services			Pilot	Pilot	Publication
Land-based & Environmental Industries					Pilot
Performing Arts & Entertainment Industries					Pilot

Foundation	1992–93	1993–94	1994–95	1995–96	1996–97
Art & Design		Pilot	Publication	Revised pubn	
Business		Pilot	Publication	Revised pubn	
Health & Social Care		Pilot	Publication	Revised pubn	
Leisure & Tourism		Pilot	Publication	Revised pubn	
Manufacturing		Pilot	Publication	Revised pubn	
Construction & the Built Environment			Pilot		Publication
Hospitality & Catering			Pilot		Publication
Science			Pilot		Publication
Engineering			Pilot	Publication	
Information Technology			Pilot	Publication	
Media: Communication & Production					(Pilot)*
Retail & Distributive Services					(Pilot)*
Land-based & Environmental Industries					(Pilot)*
Performing Arts & Entertainment Industries					(Pilot)*

*Possible introduction

Figure 11.2 Phased introduction of GNVQ vocational areas

registered for GNVQs—almost one in seven of all 16-year-olds—including 14 500 studying leisure and tourism, and 1800 studying manufacturing. For those students with poor GCSE results who plan to retake these examinations post-16, the prospect of taking a level 2 GNVQ Intermediate course (equivalent to four GCSEs at C or above) is an attractive one; certainly a 'more stimulating and useful option than the dispiriting option of resitting GCSEs without necessarily doing better a second time' (Kingston, 1994). This potential market is an exciting one for geography education—a point not missed by the Geographical Association's *Response to the Dearing Consultation* which stated 'we believe that the geography input into GNVQ could and should be greater than it currently is' (GA, 1993).

The national framework for vocational qualifications is illustrated in Figure 11.3.

The introduction of GNVQs has major implications for staff development, departmental workloads, candidate recruitment and resourcing. The initial planning for courses was supported by INSET materials, videos and documentation from the awarding bodies and the NCVQ to support the changes required in teaching and learning activities, administration, record keeping, testing and assessment. However, as Marvell (1994) observed, the reaction of geography departments depended greatly on the experience and background of its individual staff. Geography teachers were increasingly expected to have some relevant vocational experience, and be willing to co-operate with other departments in the delivery of courses. The NCVQ did not underestimate the magnitude of the changes that this required for it advised that 'Only schools experienced in vocational education, which have staff with appropriate subject expertise, suitable resources and contacts with industry should consider offering GNVQs at this time' (NCVQ, 1994, p. 9).

NVQs are more closely linked to the workplace, which means that their impact on Key Stage 4 and the 16–19 age range in full-time education may be less than GNVQs—perhaps only involving those schools with established industrial links. Interestingly many schools in France, Germany and The Netherlands offer such courses and see few conflicts with their broader educational aims. In other European countries, children also have a choice of a general, technical or vocational curriculum, rather than individual subjects, during their final years at school.

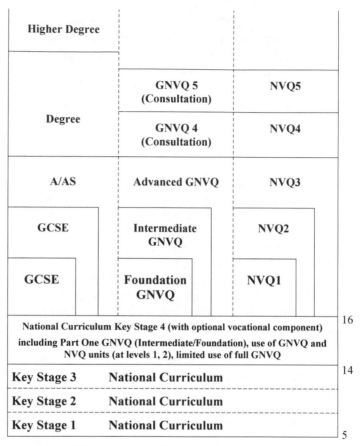

Figure 11.3   National qualifications framework

GNVQ—the link to universities

There was concern that universities would view GNVQs as wholly vocational qualifications with little relevance to fulfilling entry requirements to higher education degree or diploma courses. Nine hundred candidates taking the Advanced GNVQ applied to HEIs in 1994—but with many institutions demanding Advanced GNVQ at 'merit' or 'distinction' grades, or in combination with specific grades at A or AS level, straight GNVQ entrance appeared to be some distance away (Meikle, 1994). Universities feared that GNVQ applicants would have educational weaknesses, especially with regard to subject knowledge, that required additional teaching—a point reflected in Smithers' (1993)

Dispatches Report (Channel 4, 15 December 1993). Two critical OFSTED reports (1993, 1994) on the introduction of GNVQs into schools also highlighted and extended some of these concerns. The later report called for 'rigorous and more manageable assessment, better course design and tougher external checks' to be applied (OFSTED news release, 31.10.94). In addition the achievement of consistent assessment standards both within schools and on a national scale was questioned. The Advanced level GNVQ was discovered to be broadly on a par with two passes at GCE A level, but the standards of the Intermediate level were much more variable.

Many of these fears were restated by the Further Education Funding Council (FEFC, 1994) inspectors, who had 'major concerns' about the teaching and assessment of higher level GNVQs, the lack of employer involvement in courses and the extent of documentation overload (Utley, 1994). Pressure was further increased by a CBI report (1994) and the action of a network of educationists calling themselves 'Article 26'.

The move towards offering vocational courses up to degree level at universities may be inevitable, despite resistance among academics at some of the older establishments. Indeed many of the former polytechnics, as well as the Open University, already offer work-based NVQs to students. According to how the universities respond to the challenge, we may see an increasing balance of vocational and academic courses offered in all institutions, or alternatively the return to some form of the old 'university—polytechnic' divide.

Conclusions

Geographers have always sought to provide children with 'relevant' forms of education. However, as schools and colleges took up the vocational challenge in the 1990s the qualifications designed to promote such relevance were being widely criticised (CBI, 1994; OFSTED, 1993, 1994; FEFC, 1994; Smithers, 1993; Utley, 1994; Targett and Tysome 1994a, b). Confusion existed over whether schools should consider providing narrower vocational training and education closely linked to the 'world of work'; or whether a more vocational emphasis from within traditional subjects was required.

With increased 'staying on' rates post-16, very few children faced entering the employment market as untrained or poorly educated 'factory fodder'—the major issue was what form this education and training should take. Employers, for example, called for schools to provide general business awareness, literacy, numeracy and IT skills rather than specific forms of job training (CBI, 1994). Geography has a role to play here, for the subject can act as the foundation for a variety of vocational courses and help to create an understanding of the wider world. Because of this employers have traditionally welcomed geographers, for they often possess skills, knowledge and understanding relevant to the work environment.

Interestingly, Walford (1981) sounds a cautionary, if somewhat pessimistic, note. In his account of the 'utilitarian tradition' in geography education he states that although subjects with clear vocational aims are valued, geography does not rank highly as one of these. Thus 'the role of geography, seen within this tradition, will be defined in an adjunctive role; it will be informational and descriptive to some extent, designed to equip students to read TV bulletins without being bemused, and to maintain casual conversations, but not to *be* geographers' (Walford, 1981, p. 220).

Many students' decisions about whether to study GNVQs, A/AS levels or some combination of all three is influenced by their perception of each qualification's acceptability for admission to higher education. A clarification of the roles of all 16–19 qualifications is expected from the second Dearing Review which began in 1995. Teachers and students need to have some idea of how admissions tutors at universities 'value' different vocational and academic awards; for example, despite their reasonably warm welcome in 1987 AS levels made a slow start as an entry qualification to higher education. Meikle (1994, p. 4) has warned that because universities are being threatened with financial penalties if they over-recruit many are currently tempted 'to play safe' and favour applicants with A levels—a qualification which they understand and trust—rather than experimenting with GNVQ entrance. Alternatively, some universities may seek the combination of GNVQs with A or AS levels at specific grades.

In the mid-1990s it was difficult to avoid the conclusion that the whole 14–19 educational system was in a confused state. 'Base GNVQs (appeared to be) losing the alchemical struggle to become golden A levels' (*THES*, 'Opinion', 4.11.94) as employers and HEIs showed inconsistency in their acceptance of the former. It was also clear that although the previously poor record of low staying-on rates post-16 had finally been broken, the quality of many of the courses and qualifications that students now took was questionable. The creation of a credible, unified, academic and vocational education system with nationally agreed standards was not served by a government keen to 'shore up' A levels and tinker with vocationalism. Both A levels and GNVQs suffered at this stage from inconsistencies—among the examination boards in the case of the former, and due to variations in regional standards in the case of the latter. Modular courses in both, including externally assessed core units, might have gone some way towards establishing the standardisation that was desperately needed across the academic/vocational divide.

In future it may become common that those students who take geography as a single or joint honours degree at university will not necessarily have studied it at school beyond the age of 14. Additionally the growth of modular courses at university creates the flexibility for students to 'major' in geography at some later stage of their higher education programme. This could be the case with GNVQ entrants who have 'geography related' GNVQs (or a GNVQ plus A level), but who only study geography within higher education after a first-year foundation course. In some ways this development may not be radical, for even in the past, universities could not guarantee the geographical knowledge, skills and understanding of their intake due to variations in A and AS level geography syllabuses.

Additionally other subjects (such as philosophy, law, politics) have all survived in higher education without a direct 'feeder' subject in schools, therefore why should geography be any different?

This situation may create for some observers the perception of a 'hierarchy' of different types of geography education and geography students: the 'élite' students being those who have studied geography GCSE, geography A level, and single honours geography; while the 'non-élite' have taken GCSE/GNVQs in 'geography-related' areas of study, GNVQs and/or A/AS levels, eventually leading to joint honours in geography.

So what of the future? The merger, announced in June

1995, between the Departments of Employment and Education will almost certainly give added momentum to moves to bring academic and vocational 'pathways' closer together. In addition, by the year 2000 the revised National Targets for Education and Training, supported by the government and a wide range of commercial organisations, expect 85 per cent of 21-year-olds to achieve at least two A levels (or the equivalent GNVQ level 3); while 85 per cent of 19-year-olds should be gaining five GCSEs at grades A to C. These targets are considerably short of the actual figures for 1993 of 37 and 61 per cent respectively (Nash and Dean, 1994). This proposed expansion within the educational market, and the implied raising of standards, is therefore marked—a situation in which geography education must find both an academic and a vocational niche.

As Rawling (1994, p. 99) states, our main advantage in attempting to achieve this goal 'may lie in geography's proven potential to contribute to both vocational and academic studies, as long as we get this message clearly across'.

References

Bennetts, T. (1994) The Dearing Report and its implications for geography. *Teaching Geography*, **19**(2), 60–63.

Blackburne, L. and Nash, I. (1994) Breakthrough welcomed. *TES*, June 24. 7. 94.

Butt, G. (1993) The scenario post 16. *Teaching Geography*, **18**(1), 36–37.

Butt, G. and Lambert, D. (1993) Modules, cores and the new A/AS levels. *Teaching Geography*, **18**(4), 180–181.

CBI (1994) *Quality Assured: The CBI Review of NVQs and SVQs*. CBI, London.

Chorley, R. and Haggett, P. (1965) *Frontiers in Geography Teaching*. Methuen, London.

Dearing, R. (1993a) *The National Curriculum and its Assessment. Interim Report*. SCAA, London.

Dearing, R. (1993b) *The National Curriculum and its Assessment. Final Report*. SCAA, London.

Dearing, R. (1995) *Review of 16–19 Qualifications: Summary of the Interim Report; The Issues for Consideration*. Central Office for Information, London.

DES/WO (Department of Education and Science and Welsh Office) and Department of Employment (1991) *Education and Training for the 21st Century*. HMSO, London.

DfE/WO (Department of Education and Welsh Office) (1994) *Competitiveness: Forging Ahead*. HMSO, London.

FEFC (1994) *General National Vocational Qualifications in the FE sector in England.* FEFC, London.

GA (1993) *Response to the Dearing Consultation,* November: Sheffield.

GCE Boards of England, Wales and Northern Ireland (1983) *Common Cores at Advanced Levels.*

GNVQ (1993) *GNVQ Information Note,* September. NCVQ, London.

GNVQ (1994) *GNVQ Newsletter.* Issue 2, February. NCVQ, London.

Kingston, P. (1994) A new lease of life. *Guardian Education,* 14 June.

Marvell, A. (1994) Should your department consider GNVQ leisure and tourism? *Teaching Geography,* **19**(4), 175–176.

Meikle, J. (1994) Parlez-vous GNVQ? *Guardian Education,* 14 June.

Nash, I. (1994) Vocational revolution in trouble. *TES,* 21 October.

Nash, I. and Dean, C. (1994) Whitehall sets tough targets on exams. *TES,* 15 July.

OFSTED (1993) *GNVQs in Schools,* October. HMSO, London.

OFSTED (1994) *GNVQs in Schools,* October. HMSO, London.

Orrell, K. and Wilson, P. (1994) GCSE geography in the post-Dearing era. *Teaching Geography,* **19**(2), 90–91.

Pyke, N., Blackburne, L. and Nash, I. (1994) A levels at risk in drive to compete. *TES,* 3 June.

Rawling, E. (1994) Dearing and the National Curriculum: what next for geography? In R. Walford and P. Machon (eds), *Challenging Times: Implementing the National Curriculum in Geography,* Longman, Harlow, pp. 98–99.

RSA (1994) *GNVQ Centre Guidelines: a Guide for Centres in the Delivery of General National Vocational Qualifications.* RSA Examinations Board, Coventry.

SCAA (1993) *GCE A and AS Subject Core for Geography.* SCAA, London.

SCAA (1994) *Geography in the National Curriculum. Draft Proposals.* HMSO.

SCAA (1995) *Using the AS.* SCAA, London.

SEAC (1992) *A and AS: Principles for GCE Advanced and Advanced Supplementary Examinations.* SEAC.

Smithers, A. (1993) *All Our Futures: Britain's Education Revolution: A Dispatches Report on Education.* Channel 4 Television, London.

Swatridge, C. (1994) Stalling AS may blow up the engine this time. *TES,* 3 June.

Targett, S. and Tysome, T. (1994a) Whistle blowers under fire. *THES,* 24 June.

Targett, S. and Tysome, T. (1994b) CBI insists on reformed vocational skills system. *THES,* 24 June.

Utley, A. (1994) Teething troubles plague GNVQs. *THES,* 24 June.

Walford, R. (1981) Language, ideology and geography teaching. In R. Walford (ed.), *Signposts for Geography Teaching,* Longman, Harlow, pp. 215–222.

Wilson, P. (1994) A level analysis. Nutshell notes. *Geographical Magazine,* **58**, March.

Defining and measuring progression in geographical education

12

Richard Daugherty

Introduction

The idea of progression is implicit in any discussion of the nature of the learning we hope students will engage in. If we did not hope that our students would, in some sense, progress we would have no foundation on which to construct a curriculum or to embark on the act of teaching. Indeed the rewards of teaching are typically expressed in terms of the responses of learners, whether it be an unexpected insight offered by a previously reluctant learner or the 'glittering prizes' awarded to the high-flier. However, while any teacher of geography can identify such instances from his/her experience, it is quite another matter to attempt to develop a systematic view of the ways in which the curriculum for a group of geography students might make provision for those students to progress. It is an even more daunting prospect to look across the whole of the period of formal geography teaching in schools and colleges and to ask what concept of progression, implicit or explicit, is informing the design of the curriculum.

The need to confront the issue of progression in school geography became urgent when, in 1988, schools in England and Wales were faced with the prospect of implementing the various requirements of the Education Reform Act, including the introduction of a National Curriculum and its

Geography into the Twenty-first Century, Edited by E.M. Rawling and R.A. Daugherty,
© 1996 John Wiley & Sons Ltd.

associated provision for national assessment of pupils aged 7, 11, 14 and 16. The challenge was initially one for those at the centre who were given the responsibility for designing a curriculum framework for geography. Once that framework was in place, it was for teachers to modify their syllabuses, schemes of work and classroom practice to take account of the new requirements and to try to turn the hoped-for progression into a reality for their students.

The demands of a detailed curriculum specification, the development of which is discussed elsewhere (see Rawling, p. 248) were challenging enough. But perhaps the most difficult features for curriculum designers and teachers to interpret, and to give effect to, were those arising from the assessment model on which the government chose to build the new system. That model could have been constructed on the familiar basis of ranking pupils according to the marks awarded in age-specific national tests for each age-group. Instead, the Task Group on Assessment and Testing (TGAT) (DES/WO, 1988), set up by the government to advise on how to introduce national assessment, opted for the very much more ambitious idea of defining specific attainments in each subject, including geography (Daugherty, 1995).

The specifying of prospective learning outcomes was a matter on which geographers already had at least 15 years' experience to base recommendations for the National Curriculum. But what was outside the experience of most geography teachers and educationists were proposals that:

1. The attainments should be classified into subsets in each subject, subsets which came to be known as 'attainment targets'.
2. Within each attainment target the 'statements of attainment' should be placed at one of 10 levels, the sequence of which would describe the broad progression of attainment in that aspect of the subject.
3. Assessments of each pupil's attainment would be devised which would make it possible not only to measure progress but also to report it in terms of the specific attainments defined initially.

Thus those with responsibility for designing the National Curriculum Geography framework were faced, within the short period allowed for planning the new system, with reaching agreement on:

1. *Identifying* those aspects of the subject to be designated as attainment targets.
2. *Defining* a limited number of statements of attainment which, taken together, would represent the most significant learning outcomes of pupils studying the subject from the age of 5 to the age of 16.
3. *Structuring* those attainment statements across the 10 levels in each attainment target so that they offered a basis for tracking the progress made by a pupil and associating each of the statements with one of the attainment targets.
4. Deciding on methods of assessment which would make possible the *measuring* of each pupil's progress in relation to that framework of targets and levels.

While there was some relevant experience to draw on, time was short and the opportunities for trialling proposals were minimal given the tight timetable set by a government intent on introducing the National Curriculum as quickly as possible.

Defining learning outcomes

The trend towards using specific objectives in curriculum planning, promoted in the USA by curriculum theorists such as Mager (1962), gradually took root in the flowering of curriculum development initiatives in England and Wales in the late 1960s and early 1970s. Though objectives-based planning was, and remains, an approach criticised by some as having a potentially damaging effect on curriculum and on pedagogy (see e.g. Stenhouse, 1975), such criticism did little to slow the trend towards redefining the school curriculum, in geography as in other subjects, in such terms. In the major Schools Council sponsored curriculum projects of the 1970s it became normal to indicate the learning objectives associated with a unit of work, typically under the subheadings of 'knowledge', 'understanding', 'skills' and 'values and attitudes'.

One of the earliest and most influential attempts to define attainment objectives across the whole subject, not just for one aspect of the subject, was Her Majesty's Inspectorate (HMI) booklet, *The Teaching of Ideas in Geography* (1978). Strictly speaking what the HMI who wrote that pamphlet had produced was not a list of objectives but rather some suggestions as to *generalisations* which might underpin the learning of geography in each of its systematic subdivisions. However, it would only have taken the addition of wording such as 'pupils should be able to demonstrate an understanding of . . .' to turn those generalisations into a set of learning objectives to be included under the 'understanding' section of the conventional fourfold classification of types of objective referred to above.

Some of the problematic features of any such list are well illustrated by the way in which geographical understanding is represented in *The Teaching of Ideas*. In all there are 146 generalisations listed, some of them very lengthy in themselves, one having 9 subdivisions within it. They are grouped into 11 categories such as 'weather and climate' and 'manufacturing industries' though the authors disclaim any assumption of comprehensiveness in their coverage: 'it is not suggested that the categories used are exhaustive' (HMI, 1978, p. 2). While their interesting proposals only related to the 'understanding' type of objective (though an appendix detailing mapwork objectives did include reference to map skills), the HMI were illustrating one of the recurring themes across all attempts to define attainment in geography, or indeed in any other broad field of learning. It is not easy to select a manageable number of attainment statements within a small number of domains (or, in National Curriculum terminology, 'attainment targets').

During the late 1970s and the 1980s other authors sought to define the attainments characteristic of geographical learning. In his book, *Graphicacy in Geography Teaching*, published in 1983, Boardman set out 100 'graphicacy skills' in the form of what a pupil of a given age (7, 9, 11, 13, 19) 'should normally be able to do'. On a broader front, in the Geographical Association's *A Case for Geography* (Bailey and Binns, 1987), a working party chaired by Wiegand laid emphasis on the experiences of geographical learning 'which most pupils should have' by the ages of 7, 11, 14 and 16. Wiegand and his colleagues also used the conventional categories of 'knowledge and understanding', 'skills' and 'values and attitudes' to suggest expectations for 14- and 16-year-olds. In their focus on four age groups the working party was foreshadowing the National Curriculum key stages but, like all those who devised lists of learning outcomes in the years before 1988, it did not attempt to express these as attainments which could apply to a pupil at any age between 5 and 16.

Period	Stage	Mental age range (years)
Sensor-motor	I Sensor-motor	0–2
Preparation for and use of,	II Pre-operational	
concrete operations	A. Preconceptual	2–4
('latency')	B. Intuitive	4–7
	III Concrete operations	7–11.5
Formal operations	IV Formal operations	11.5–

Figure 12.1 Piaget's theory of stages of development. *Source:* Childs (1973)

Defining progression

The idea of progression in any kind of learning is something which can be understood experientially, by reference to the way in which a particular learner feels him- or herself to be gaining mastery over something. But it has proved much more difficult to develop schema which provide us with a satisfactory overview of the ways in which learners typically progress. Indeed it is arguable that learning is so personal and idiosyncratic that there can be no such overview defining commonality in the experiences of all learners.

One of the theoretical constructs which has proved most influential in guiding the thinking of curriculum planners is the notion of the 'stages of development' which Piaget proposed, based on a wealth of experimental evidence (Figure 12.1). Naish (1982, p. 31) drew attention to some of the implications of Piaget's ideas for geographical learning. For example:

> It would seem appropriate to provide opportunities for accurate and thorough descriptive work in the concrete operational stage [ages 7 to $11\frac{1}{2}$], leading on to problem-solving in the later stages.

But other writers have challenged the relevance of Piaget's findings to curriculum planning. Egan (1979), for example, points out that Piaget's schema is concerned only with the child's intellectual development and he questions the wisdom of basing progression solely on a type of model which he

Stage	Age range (years)
Mythic	4/5–9/10
Romantic	8/9–14/15
Philosophic	14/15–19/20
Ironic	19/20–

Figure 12.2 Egan's theory of educational development. *Source:* Egan (1979)

refers to as 'genetic epistemology'. In its place Egan puts forward, without any research evidence at all, a model of 'educational development' (Figure. 12.2).

Egan's argument contains several examples of how a particular theoretical basis, implicit or explicit, adopted for planning a curriculum could make a substantial difference to the type of experience available to learners at a particular age/stage (Egan, 1979, p. 47):

> . . . all curricula that are composed on the assumption that students' knowledge of the world must be gradually extended on lines of content association from the self and local experiences, perform an enormous disservice to students' educational development. They will also, incidentally, bore most students out of their minds.
>
> It is perhaps not surprising that this ['Romantic'] stage of intellectual wonder and excitement is also the stage of most acute boredom. If the mind is not caught up and flying in wonderful realms, it has to descend into the everyday world against which it has developed little conceptual defence.

Has the geography curriculum, typically working out from the familiar to the unfamiliar in a way which accords with Piaget's 'concrete operations' stage, been ignoring the potential for stimulating students at Egan's 'Romantic' stage (from 8/9 to 14/15) to expand their geographical and intellectual horizons by exploring what is outside their experience and relating that back to what they know?

Such theorising about progression provides us with an interesting backcloth to, rather than a firm foundation for, the task of conceptualising progression in geographical learning. Writers on progression in geography have tended to confine themselves to broad assertions about what characterises such progression. *The Teaching of Ideas in Geography*, referred to above, argues that (HMI, 1978, p. 11):

> Although what can be achieved at a particular stage
> will depend largely upon pupils' experience and their
> intellectual capabilities, important ideas need to be
> returned to time and again, with a gradual extension
> and deepening of understanding.

The authors go on to suggest a number of considerations which should be kept in mind when planning for progression in understanding, namely distance of understanding, degree of complexity, degree of abstraction, degree of precision, the involvement of attitudes and values. The genealogy of that list of considerations can be traced through Bennetts' 1981 article and the HMI paper *Geography from 5 to 16* (DES, 1986) to the Final Report of the National Curriculum Geography Working Group which argued (DES/WO, 1990, p. 12):

We envisage progression occurring in:

(a)   the breadth of studies . . .;
(b)   the scale of the area studied . . .;
(c)   the complexity of the phenomena studied and the tasks set;
(d)   the use of generalised knowledge and abstract ideas;
(e)   the precision required in practical and intellectual tasks; and
(f)   awareness and understanding of social, political and environmental issues involving different attitudes and values.

While identifying some of the dimensions which may characterise progressive mastery of the subject, this list does not pretend to offer a complete set of parameters to be taken into account when designing progression into a curriculum. Indeed the list may be more useful in guiding the choice of the specific topics and places to be included in the day-to-day planning of appropriately differentiated work for a particular teaching group. The HMI pamphlet (DES, 1986) did include as an appendix an illustration of how considerations such as those referred to above could be taken into account when designing a series of units of work on manufacturing industry. Other similar, necessarily tentative, structurings of learning content and outcomes on a progressive basis can also be found in the literature of the 1980s.

It is only in the field of map learning that a sufficient body of empirical research was available for suggestions as to appropriate learning sequences which go beyond what any geography teacher could propose by drawing on personal experience. Basing their conclusions on evidence from 12 writers and researchers on map learning, Gerber and Wilson (1984) proposed a mapping skills sequence, arguing that:

> . . . a suitable sequence to introduce the essential skills for mapping may be developed using four of the key properties of a map, namely:
>
> 1.   Plan View
> 2.   Arrangement
> 3.   Proportion
> 4.   Map Language.

Other researchers in this field, including Catling (1979, 1980) and Boardman (1983), have also used research findings as the starting-point for proposing appropriate sequences for a map skills curriculum. However, though such sequences may assist in curriculum planning, they do not in themselves constitute a curriculum framework. Other considerations, such as course structure, content, context and pupil motivation, have to be taken into account for map skills to become an integral part of a geography curriculum (see Graves, 1979).

Measuring attainment

In the pre-National Curriculum era there were no established procedures in place for us to assess pupils' progress over time. All the main national systems for assessing geographical learning had two main characteristics. Firstly, they were geared to taking a 'snapshot' of attainment at a specific point, typically being targeted at pupils of a given age, such as GCE O level or CSE for 16-year-olds and GCE A level for 18-year-olds. The way the requirements for those age-related assessments were set out—course content, course objectives, grading of outcomes—made it impossible to relate an individual's performance on one examination *directly* to what she/he had achieved earlier in her/his geographical education. The second main characteristic was that the results of those assessments, typically a combination of end-of-course examinations and 'coursework', were expressed in terms of overall grades, revealing nothing of the specific qualities the pupil had demonstrated in achieving a particular grade.

It is important to recognise the distinction between schemes of assessment which use explicit attainment criteria in the design of tests or to aid the allocation of marks and those schemes of assessment which seek not only to do that but also to make criteria-based judgements as to whether a pupil has demonstrated a certain attainment and then to report outcomes in those terms. Most assessment initiatives in geographical education in the 1970s and 1980s were of the former type, i.e. they were explicit about attainment criteria but then judged performance in terms of marks and grades. Thus a pupil taking an Avery Hill GCSE geography course would be following a syllabus with specified assessment objectives, such as 'to describe and offer explanations of spatial patterns and relationships'. Her/his coursework would be marked using detailed criteria for allocating marks. Yet the result of the examination would be reported as a grade such as B or E.

Only a very few attempted to adopt the alternative approach, criterion-referenced assessment, involving the reporting of results in terms of particular, prespecified profiles of attainment. Even fewer met success except where the results were to be used only within the school for feeding back information to the pupil and his/her parents. And yet it is only when what we are hoping pupils will learn is spelled out that it becomes possible to envisage measuring and reporting progression in relation to those skills and ideas we have defined as symptomatic of geographical ability. The measurement of progression, in geography as in any other subject, depends in the first instance on defining attainments in such a way that measurement of progress can at least be attempted.

Some individual secondary school geography departments embarked upon that task in the 1980s, typically prompted by the need to track and record a profile of attainment in the subject for inclusion in records of achievement (Graves and Naish, 1986). Another initiative was the research and development project on 'diagnostic assessment in geography' (Black and Dockrell, 1980) to which a number of Scottish schools contributed in the late 1970s and early 1980s. A further attempt to take attainment criteria through the marking process and into the reporting of outcomes has also been in Scotland, with the use of 'grade related criteria' in the Standard Grade examination (see Hunter).

A similar initiative appeared likely to come to fruition in England and Wales with the introduction of the new General Certificate of Secondary Education (GCSE) for 16-year-olds in 1986. Working parties were established in 10 major subjects, including geography, with a view to defining levels of attainment which could then be measured in the examination and reported to the 'user' of the examination result. It is instructive, in view of the later attempt to do something similar for National Curriculum assessment, to ask why the idea of measuring and reporting specific attainments ran into insuperable problems and was subsequently abandoned.

The Working Party looking at GCSE geography (SEC, 1985) recommended five 'domains':

1. Specific geographical knowledge;
2. Geographical understanding;
3. Map and graphic skills;
4. Application of geography to economic, environmental, political and social issues;
5. Geographical enquiry.

In each of these there were to be four levels of attainment, with lists of learning outcomes set out for each domain at each level (though the Report only put forward proposals for three of the levels).

In effect, there was a five-by-four matrix which would, in the first instance, be used to influence the design of examinations and coursework and later would form the basis on which candidate performance would be reported. There would be a profile of attainment in each domain and an aggregation across the domains, according to certain rules, to produce an overall subject grade. If this had been done, GCSE geography would have become a criterion-referenced examination. In the event, while many of the requirements of the examination are spelled out in terms of explicit attainment criteria, GCSE awarding still depends mainly on the rank order of candidates arrived at by totalling the marks on each question. It remains therefore essentially a norm-referenced process with the levels of award being determined largely by marks gained rather than by the matching of each individual's performance to prespecified descriptors of geographical attainment.

Some of the problems inherent in defining and measuring attainment are touched upon in the Grade Criteria Working Party's Report. In respect of geographical knowledge, is it possible to conceive of progression as distinct from increasing quantity of knowledge? More broadly, are the proposed domains distinct enough from each other to make possible, firstly, the design of tasks which clearly relate only to that domain and then, later, the reporting of the results for each domain separately? The Report also hinted at some of the problems which were likely to be encountered should the proposals ever be implemented, though the Working Party's brief had excluded, as the brief of the National Curriculum Geography Working Party was to do four years later, any serious consideration of the practicalities of actually assessing the attainment framework being proposed.

Discussion of, and preliminary development work on, the geography grade criteria took place but, after several months of trials, the government abandoned the proposal to use explicit criteria as the basis for measuring and reporting attainment in GCSE. It was an unpromising augury for the subsequent attempt to do much the same thing, not just for one age group following specified syllabuses but for all pupils across the full age range of compulsory schooling from 5 to 16.

Progression in National Curriculum Geography

How much advance has there been during the first five years of National Curriculum Geography in defining learning outcomes, defining progression and measuring whether the learner has indeed 'progressed'?

The first phase of work in this respect was the responsibility of the Working Group appointed by the Secretaries of State to bring forward proposals. The results of that work were published in June 1990 as *Geography for Ages 5 to 16* (DES/WO, 1990). The subsequent Statutory Orders (DES, 1991; WO, 1991) which would form the basis for implementing National Curriculum Geography differed in important matters of detail from the Working Group's recommendations. But the Group's approach to progression was unaffected by those changes. That approach had been expressed in the Group's Report (DES/WO, 1990, p.12) in terms very similar to the six aspects which had been listed by HMI in 1986: breadth, scale, complexity, abstraction, precision, awareness of issues. It was, however, difficult to discern how such thinking had been applied in the process of structuring and sequencing the statements of attainment in the attainment targets. The problem of articulating aspects of progression in each of the 14 proposed 'strands' running through the levels of attainment was all too evident when it came to stipulating the statements of attainment across 10 levels in the 7 (later reduced to 5) attainment targets.

How to express an intended outcome of learning in the form of a statement of attainment? For example, pupils at level 7 should 'be able to analyse the processes which lead to changes in towns and cities, and their effects'. But which processes leading to which changes in which towns and cities? And what kind of performance by the learner would be regarded as a successful analysis? The well-established difficulties encountered in all attempts to define a curriculum in criterion terms were revealing themselves. Greater specificity results in hundreds or perhaps thousands of separate 'bits' of geographical attainment. On the other hand, limiting the number of statements brings with it the problem of loose, vague descriptors which are all but impossible to interpret in a consistent way. For earlier advocates of criterion referencing (e.g. Popham, 1978) the answer lay in very detailed specifications, setting out the context in which attainment would be judged and the criteria to be applied in judging mastery. For those seeking to draw up a National Curriculum which did justice to all aspects of the subject such an encyclopaedic approach to the task was never a realistic, nor even a defensible, proposition. Vagueness in defining learning outcomes would be an inherent feature of any attempt to provide a framework for geographical learning for all age-groups and ability levels.

If defining learning outcomes was difficult, how much more problematic would be the task of ordering the statements of attainment so that they offered a basis for describing progression across 10 levels? For example, in the same human geography attainment target from which the above example was taken we find '[pupils should be able to] explain the reasons for the growth of economic activities in particular locations' at level 5 and 'analyse changes in the distribution of selected economic activities, and the effects of those changes' at level 7. We can infer from this that analysis of distribution and effects is seen as being more difficult than the explanation of location. To quote this one example is not to suggest that the Working Group might have got this or any other sequence of difficulty 'wrong' but rather to emphasise the impossibility of associating a necessarily general attainment statement with a particular level of difficulty. The Group made life more difficult for itself by trying to link specific locations to particular levels in a few cases (see Rawling, p. 259), harking back to the days when the southern continents were taught to younger secondary pupils because they were supposedly simpler and easier to understand than the complexities of Europe. But that aberration was only an incidental feature of an exercise in mapping progression which lacked a firm grounding in epistemology, learning theory or curriculum design.

With the abandonment of the proposed national tests in geography, the challenge of finding ways of measuring progression was left to individual teachers seeking to measure the attainments of each of their pupils. In the course of trying to meet the perceived requirements of routine 'teacher assessment' many teachers found themselves engaged in elaborate and time-consuming exercises of assessing and then recording the details of the performance of each pupil in relation to each statement of attainment (Daugherty and Lambert, 1994). If, as is argued above, the map of progression they were working from was flawed, it is not surprising that they often found the exercise to be unproductive as well as burdensome.

If the 'route map' is indeed but a sketch, an outline of the features along the way which may (or may not) be symptomatic of progression, it should not surprise us that neither teacher nor learner has been able to follow it, ticking off each landmark as it is passed. The blame for the anxiety and wasteful effort associated with trying to measure progression in this way does not lie only with the Geography Working Group. That group did its best to draw its sketch of progression, though it seems to have conceptualised progression in terms of the sequence of topics to be taught to the pupil, an approach which focuses on what is to be studied rather than what is to be learned through studying. The attainment targets were used to reinforce the 'this must be taught' messages of the programmes of study rather than to complement the programmes of study by setting out the types of learning outcome which might be identified in pupils who had completed such programmes.

But the problems encountered by the Geography Working Group are also a consequence of the model, the TGAT model, to which they, and other equivalent groups, were working. Those responsible for establishing a national assessment system had led those designing and implementing the system to believe that it was feasible to stipulate statements of attainment allocated to each of 10 levels which would offer a valid, reliable basis for measuring how well students were progressing. As a laudable attempt to lift our sights from planning the curriculum separately for each age group, the *idea* of 10 levels of attainment was radical and a potentially powerful tool for beginning to rethink both sequencing and progression in the curriculum. But it was too much of a tentative, provisional sketch of how different attainments could be related to each other in terms of difficulty to be a sound foundation for practice in assessing and recording attainment.

The revised National Curriculum Orders (DfE, 1995; WO, 1995) acknowledged the impossibility of charting a detailed route map. Instead, it ensured that the 'this must be taught' messages were located in the programmes of study. Alongside those programmes of study was a single attainment target, conflating the key aspects of attainment in geography into eight 'level descriptions' (plus one for 'exceptional perform-ance'). By deconstructing the level descriptions it is possible to infer the thinking behind this simpler, more user-friendly route map. Thus, for example, at level 3 pupils should 'show an awareness that different places may have both similar and different characteristics', but by level 5 they should be able to 'describe how [human and physical] processes change the environment and can lead to similarities and differences between places'. Each level description is a multifaceted composite statement incorporating many distinct aspects of geographical learning.

Unlike the statements of attainment which they replaced the level descriptions do not imply that there might be a precise route map in there somewhere if only we could understand the wording and think up more accurate methods of measurement. They are self-evidently 'broad brush' impressions to be used by teachers in arriving at rough estimates of the performance of each student. Whether such estimates are going to be made in a sufficiently consistent way by different teachers to have any real meaning or value remains to be seen. SCAA and ACAC publications during 1996 will exemplify the use of level descriptions in reaching judgements about the quality of pupils' work. At this early stage of a new simpler, slimmer framework we can at least be grateful that the official documents are now recognising that we do not have sufficient understanding of how to define progression, still less of how to measure it, to be able to attempt anything more than a rough estimate.

In the original design of the National Curriculum, what appears to have been aspired to was a template of attainment, in geography as in other subjects, which could be placed over the work of an individual pupil to give us a precise picture of his/her attainment. The prospect of an all-purpose, all-inclusive system for defining and measuring attainment proved too attractive for those who put forward the original design brief on which the National Curriculum was based. If the originators of the system had taken account of prior experience of defining learning outcomes and charting progression they might have been forewarned as to the foolhardiness of such an enterprise. Only in the most limited and circumscribed areas of skills learning, where context and assessment are tightly controlled, has it proved possible to approach progression in this way. Sadly, some of the same lessons are again being relearned in the more recent initiative of the General National Vocational Qualification (GNVQ) (Wolf, 1994).

Progression in geographical education: future directions

If the recent experience of attempting to chart progression in geography leads us to be cautious in seeking alternatives, what kind of advances could be made in this respect as we seek to capitalise on the fact of geography being a part of the experience of every pupil in a state school through to the age of 14 and for many for several years beyond that? Can we do better than simply reverting to hunches about the kind of learning which is appropriate for a 7-year-old or a 14-year-old? Inherent in the very notion of a National Curriculum from 5 to 14 is the assumption that learners will progress, and the revised Geography Orders offer a variety of hints, some more obvious than others, as to how the older pupil's curriculum might be seen as extending and building on what has gone before. If we are to avoid the worst of the former world of age-related syllabuses, with the repetition and lack of development which has been all too common, we will need to continue to reflect upon, and to investigate, what it means to progress in learning geography. That challenge can be interpreted and acted upon in terms of epistemology, of pedagogy, of curriculum design and of assessment practice.

The epistemological issues are, perhaps more than any others, matters which can benefit from the interplay of academic geographers and those whose work is focused on the young learner in school. All of the attempts to define attainment in geography which were reviewed earlier in this chapter have necessarily taken a view of the nature of the subject and its component elements. The framing of those elements in terms such as 'physical geography' or 'environmental geography' rests on assumptions which often remain implicit rather than being fully articulated and justified. When a case is made for an alternative formulation (see, for example, Hall's discussion of innovative approaches to the design of A level syllabuses, p. 166) the most fundamental of epistemological questions are at issue.

There is continuing scope for exploring the epistemology of school geography, an exploration which should encompass not just the framework of concepts on which geographical learning depends but also how, looked at from an epistemological starting-point, progression might be defined. Curriculum design models such as that developed by Biddle in the 1970s (summarised in Graves, 1975, pp. 109–113) still have much to offer in relating issues of epistemology to the process of curriculum design. A rational approach to curriculum planning seems to have fallen into disfavour among the policy-makers of the late 1980s. Yet such an approach would help clarify the assumptions implicit in the pragmatic decisions about the subject which are built into the present specification of the Geography National Curriculum and it would provide us with one source of ideas when it comes to redesigning that specification.

A clearer view of the subject as a vehicle for the education of young people is not in itself an answer to the problem of planning an appropriate geography curriculum for our schools. As Ruthven (1995, p. 10) comments in his critique of the TGAT model in the context of mathematics:

> Curricular objectives specified in terms of formal epistemic analysis of mathematics take no account of the cognitive strategies which pupils actually use to tackle mathematical tasks; of the factors influencing these strategies; and of the ways in which these strategies mature and change.

What do we know of the learning of geography beyond the understanding that each of us has accumulated from the experience of teaching the subject? There is a paucity of empirical evidence derived from studies of how children learn geography. In the absence of such studies we fall back on our instincts as educators and on the theories which colour our thinking (see Piaget, Egan earlier in this chapter) to supply us with the assumptions on which our curriculum plans are built.

To take just one example, how do young people actually develop the 'awareness and understanding of social, political and environmental issues' which was proposed by the National Curriculum Working Group as one of the dimensions of progression in geography? The level descriptions of the revised Geography Orders for England (DfE, 1995) point to some ways in which such awareness and understanding could be recognised in the work of pupils at different levels. For example, while pupils at level 2 are expected to be able to 'express views on attractive and unattractive features of the environment of a locality' those at level 7 would 'recognise that human actions may have unintended environmental consequences and that change sometimes leads to conflict' (DfE, 1995, pp. 18, 19).

But how much empirical evidence can we draw on of young people and their progression in this sense, or in any other aspect of their geographical learning? Graves (1975) offered a useful overview of the research then available on children's learning of geography, but it is disappointing to note how sparse are the studies which have been undertaken since then. It is not that any amount of empirical research could be expected to provide 'answers' to complex questions of cognitive development; the substantial body of such work in the fields of mathematics and science education is testimony to that. But even a modest series of investigations of children's responses to studying unfamiliar environments would help us decide whether to construct that aspect of the curriculum on the theoretical assumptions embodied in Egan's model of 'educational development' or to continue, as we have typically done in the past, to extend pupils' knowledge of the world outwards on lines of content association from the learner and his/her direct experience. There is an important role here for those geographers whose base is in education departments in universities but whose professional focus remains on the 5–19 age range.

The realities of progression for the learner are not, however, exposed either by exploring epistemology or through theories of cognition and educational development. Those realities are to be found in classrooms across the country as teachers plan schemes of work for their pupils and support individual learners in their struggle to progress. We can learn in this context from the experience of other subjects (Butterfield, 1993) both about the difficulty of reaching a consensus on what it means to progress in a subject (see e.g. Cooper, 1994, on history) and about the strategies we could adopt to improve our understanding of progression. The case put forward by Ruthven (1995, p. 25) for mathematics could be echoed in geography:

> . . . a continuing programme of research is needed, aimed at exploiting the results of National Curriculum assessment to influence its development. The goal of this programme should be to explore the interactions between curriculum, classroom and cognition in order to improve the specification and structuring of curriculum domains and refine the means of assessing them.

The initiative for such a programme should rest with the curriculum authorities in England (SCAA) and Wales (ACAC) which have a continuing responsibility to monitor the impact of the National Curriculum Orders on learning. But the debate as to what is revealed about geographical learning by the experience of implementation should be led by those at every level—teachers, educationists, academics—who share a commitment to enhancing the contribution of geography to the education of all young people. What is needed is the mobilisation of geography teachers at all levels to facilitate the sharing of their collective experience of progression.

Conclusion

While the inadequate and incomplete nature of the notions of progression which are informing our current thinking about the geography curriculum is all too obvious, we should not underestimate the significance of the move over the past few years to conceptualising geographical learning not as a series of essentially discrete steps but as a continuum across the full span of the learner's formal education in school and college. The very existence of a series of curriculum and assessment frameworks for those who study geography from the age of 5 through to the age of 19 poses questions which go to the heart of the subject and its role in the education system (see Rawling, p. 248). Have we thought through the geography of those frameworks? Do we understand enough about the way students learn their geography? Are we taking sufficient account of the very different curriculum cultures of the schools and colleges in which they are educated (Williams, 1994)?

Similar ambitions to develop overarching frameworks for geographical education are behind the parallel developments in the United States (NAEP, 1994; Rawling, 1995) and initiatives elsewhere in the world. They present all geography educators with the direct challenge, not previously faced, of spelling out the lines, cognitive and attitudinal, along which we hope the learner will develop. If we are realistic about what is possible, either in defining progression or in measuring it, we can explore more fully, through investigation and through experience, what we understand by progression and use that understanding to influence both curriculum frameworks and classroom practice. And even then, thankfully, the student will demonstrate that she/he has her/his own personal way of progressing (and regressing!), defying any attempt to superimpose anything more than a broad outline of what it is to become a 'better geographer'.

References

Bailey, P. and Binns, T. (1987) *A Case for Geography*. Geographical Association, Sheffield.

Bennetts, T. (1981). Progression in the geography curriculum. In R. Walford (ed.) *Signposts for Geography Teaching*. Longman, Harlow, pp. 165–185.

Black, H. and Dockrell, W. (1980) *Diagnostic Assessment in Geography.* Scottish Council for Research in Education, Edinburgh.

Boardman, D. (1983) *Graphicacy and Geography Teaching.* Croom Helm, London.

Butterfield, S. (1993). National Curriculum progression. In C. Chitty (ed.) *The National Curriculum—is it working?* Longman, Harlow, pp. 101–124.

Catling, S. (1979) Maps and cognitive maps: the young child's perception. *Geography,* **64** (4), 288–296.

Catling, S. (1980) 'Map use and objectives for map learning', *Teaching Geography,* **6**(1), 15–17.

Child, D. (1973) *Psychology and the teacher.* Holt, Rinehart and Winston, London.

Cooper, H. (1994) Historical thinking and cognitive development in the teaching of history. In H. Bourdillon (ed.) *Teaching History.* Routledge, London, pp. 101–121.

Daugherty, R. (1995) *National Curriculum Assessment: a Review of Policy 1987 to 1995.* Falmer, London.

Daugherty, R. and Lambert, D. (1994) Teacher assessment and geography in the National Curriculum. *Geography,* 79(4), 339–349.

DES (Department of Education and Science) (1986) *Geography from 5 to 16: HMI series Curriculum Matters No. 7.* HMSO, London.

DES (1991) *Geography in the National Curriculum.* DES, London.

DES/WO (Welsh Office) (1988) *National Curriculum Task Group on Assessment and Testing: a Report.* DES/WO, London.

DES/WO (1990) *Geography for Ages 5 to 16.* DES/WO, London.

DfE (Department for Education) (1995) *Geography in the National Curriculum.* DfE, London.

Egan, K. (1979) *Educational Development.* Oxford University Press, New York.

Gerber, R. and Wilson, P. (1984) Maps in the geography classroom. In J. Fien, R. Gerber and P. Wilson (eds) *The Geography Teacher's Guide to the Classroom.* Macmillan, Melbourne, pp. 146–157.

Graves, N. (1975) *Geography in Education.* Heinemann Educational, London.

Graves, N. (1979) *Curriculum Planning in Geography.* Heinemann Educational, London.

Graves, N. and Naish, M. (1986) *Profiling in Geography.* Geographical Association, Sheffield.

HMI (Her Majesty's Inspectorate) (1978) *The Teaching of Ideas in Geography.* HMSO, London.

Mager, R. (1962) *Preparing Instructional Objectives.* Fearon, Belmont, Calif.

Naish, M. (1982) Mental development and the learning of geography. In N. Graves (ed.), *New UNESCO Source Book for Geography Teaching.* Longman, Harlow, pp. 16–54.

NAEP (National Assessment of Educational Progress) (1994) *Geography Framework for the 1994 National Assessment of Educational*

Progress, report of the NAEP Geography Consensus Geography Project for the National Assessment Governing Board, Washington, DC.

Popham, J. (1978) *Criterion-referenced Measurement*. Prentice-Hall, Englewood Cliffs, NJ.

Rawling, E. (1995) *Improving the Quality of Geographical Education*. Report to the National Geographic Society, Washington, DC.

Ruthven, K. (1995) Beyond common sense: reconceptualizing National Curriculum assessment. *The Curriculum Journal*, 6(1), 5–28.

SEC (Secondary Examinations Council) (1985) *Report of Working Party: Geography Draft Grade Criteria*, SEC, London.

Stenhouse, L. (1975) *An Introduction to Curriculum Research and Development*. Heinemann Educational, London.

WO (1991) *Geography in the National Curriculum (Wales)*. WO, Cardiff.

WO (1995) *Geography in the National Curriculum (Wales)*. WO, Cardiff.

Williams, M. (1994) Progression in geographical education in the contexts of curriculum form and curriculum culture. Paper presented to the Conference of the Commission on Geographical Education of the International Geographical Union, Prague, August.

Wolf, A. (1994) *Competence-based Assessment*. Open University Press, Buckingham.

Trends in school geography and information technology

13

Sue Burkill

Introduction

During the 1990s the publication and implementation of the National Curriculum have forced teachers of geography to review all aspects of the geography curriculum. This has inevitably included a reassessment of the role of information technology (IT) in the subject.

However, before assessing the current situation it is important to remember that there is a long tradition of IT innovation by geographers (Kent, 1992). Early initiatives, particularly during the 1970s, led to a small number of geography teachers introducing IT for a range of purposes. For example, pupils were using software to reinforce basic mapwork skills, explore suitable locations for industries, investigate and project population structures and model changes in freshwater ecosystems. Statistical packages were widely used as they supported the development of quantitative techniques in sixth-form geography. This was a time in which a few well-tried subject-specific packages were used effectively by a growing number of pupils. The 1980s saw increasing support for teachers using IT in geography. The Humanities Information Technology (HIT) Project was established to fund groups of teachers supported by LEAs and higher education institutions (HEIs). The result was a number of useful publications and a network of teacher

Geography into the Twenty-first Century, Edited by E.M. Rawling and R.A. Daugherty,
© 1996 John Wiley & Sons Ltd.

groups with growing confidence in using IT. An important feature of the HIT materials was that they provided examples of the growing use of generic software by geography teachers. This opened the opportunity for IT developments tailored to a school's own needs or to the local environment. In secondary schools this trend was particularly encouraged by the publication of the *Learning Geography with Computers* pack (NCET, 1990) which combined useful software with detailed advice on classroom approaches. Many of those who were involved in developing software and disseminating ideas saw the late 1980s as a time when the use of IT was finally becoming well established in most geography departments.

More recently, it can be argued that the use of IT in geography classrooms has been both encouraged and restricted by the implementation of the National Curriculum. Some people are pointing to the positive signs. For example, many geography teachers, prompted by the requirements of the published technology Orders (DES/WO, 1990), have stopped treating IT as a luxury and have started to build it firmly into the National Curriculum framework in their schools.

On the other hand, there are also signs that the National Curriculum changes may have slowed down the diffusion of IT into geography. Teachers have had to turn their attention away from the new technologies to creating new schemes of work, rethinking resources and coping with new assessment procedures.

Given these contrasting perspectives this review comes at a vital moment for the future of IT in school geography, and it is the intention of this chapter to focus on the IT opportunities which are emerging from the National Curriculum by addressing the following questions:

1. What does the National Curriculum expect geography teachers to be doing about IT?
2. How do teachers view the integration of IT into the geography curriculum?
3. What issues need to be addressed to support an IT-enhanced geography curriculum?
4. What are the implications of an IT entitlement in school geography for higher education institutions?

Geography and IT in the National Curriculum

Revisions to the National Curriculum in England and Wales, which have taken effect from September 1995, have resulted in revised statutory requirements for both geography and IT. The revised programme of study for geography states: 'pupils should be given opportunities, where appropriate, to apply and develop their information technology (IT) capability in their study of geography' (DfE, 1995a; WO, 1995). This significantly upgrades statements in the earlier version of the National Curriculum (DES, 1991) which encouraged the use of IT without making it a statutory requirement. It is clear from the SCAA support publications, *Information Technology in the National Curriculum. The New Requirements (Key Stages 1 and 2, and Key Stage 3)* (SCAA, 1995) that IT is now a requirement in school geography courses.

There is also a well-established assumption in the IT curriculum that all subjects will be involved in implementing the IT capability statements. For example, with reference to assessment: 'There will be many opportunities for the assessment of information technology which will arise from all areas of the curriculum . . .' (SEAC, 1993, p. 9).

The revised programme of study for IT in the National Curriculum (DfE, 1995b) is important as it confirms the extent to which IT should become a normal part of a pupil's 'toolbox' of approaches to learning in geography and other subjects: 'Pupils should be taught to become critical and largely autonomous users of IT, aware of the ways in which IT tools and information sources can help them in their work' (DfE, 1995b, Key Stage 3 PoS).

At Key Stage 4, pupils should be aiming to 'work confidently and effectively with IT' in other curriculum areas as appropriate. From September 1995 geographers will, therefore, have an enhanced role to play in ensuring that pupils are familiar with and are assessed in IT skills and concepts. They need to familiarise themselves with the requirements of the IT Order and in particular with the strands of IT capability defined in the Order:

1. Communicating and handling information;
2. Controlling, measuring and modelling.

The attitudes of teachers responsible for teaching geography in primary and secondary schools will be crucial to the successful development of IT capability through the subject. It is useful to reflect on the range of attitudes or viewpoints which have emerged in the last few years towards this responsibility.

Viewpoints about the integration of IT into the geography curriculum

There are three viewpoints which coexist and deserve to be considered.

Viewpoint one: IT is mainly for enthusiasts

This first view is that using IT in geography is somewhat *eccentric* and that it lies in the realm of the *enthusiast*. This was certainly a view held widely in the past:

> At the present time the use of CAL in our discipline is still uneven and patchy both in secondary and tertiary education. Its use at a particular institution can usually be attributed to the presence of an enthusiastic member of staff, to the provision of computer resources . . . or to a combination of both (Shepherd et al., 1980, p. 21).

Given the level of investment in hardware and software in the last 15 years, it could be assumed that resources have become less of an issue (DfE, 1995c). However, staff enthusiasm still seems to be critical. Evidence to support this suggestion comes from some recent research by the ImpacT study undertaken by King's College for the DES/DfE (Watson, 1994). One feature of this two-year, longitudinal study of classes using IT was an investigation of factors which were seen to encourage or constrain the use of IT. In those geography departments where pupils were deemed to have benefited from the use of IT this was nearly always the result of the enthusiastic interest of an individual teacher. Ironically, these teachers were often labelled as eccentric or as excessively enthusiastic by colleagues who admitted to avoiding IT as a result. The dominant role of the innovator meant others could pass on their responsibilities. Many felt deskilled by confident colleagues, reacting with hostility to the coercion perceived to exist in the National Curriculum. This research revealed the importance of understanding more thoroughly the slow rate of uptake of IT by many geographers. In particular it seemed to point out the need for more sensitive long-term school-based in-service work with well-established staff. This is one area of whole school planning that needs careful consideration if the use of IT is to be more widely diffused across the curriculum.

Viewpoint two: IT is for the occasional enrichment of geography

The second view held is to regard IT as an 'added extra' which may enrich some aspects of geography, but which is not essential for teaching good geography. Teachers tend to ration the use of IT; to confine it to special occasions or to set up IT opportunities in optional sessions after school. There are good reasons for adopting this approach. IT opportunities must be set alongside the use of other enriching resources and many teachers find that, on balance, the IT resources available do not often meet their needs.

Recent statistical surveys seem to support the suggestion that this is the viewpoint of most geography teachers. In 1992, the DfE commissioned a survey of 869 primary schools and 553 secondary schools (DfE, 1993). It achieved a 78 per cent response rate. In the secondary school geography departments questioned, only 5 per cent of all geography lessons were reported to contain an IT element and only 25 per cent of departments reported that they used computers 'regularly'. However, despite this apparently low usage level, 80 per cent were using computers for some purpose (often administrative), 62 per cent were using computers in lessons at some time, and 52 per cent of geography teachers felt confident in using IT. Indirectly this survey points to IT being used sparingly in secondary classes as much because of the imposition it may place on teaching style, on organisation and on classroom management as because of a lack of courseware.

In primary schools the survey revealed that 64 per cent of the teachers used IT some time in teaching geography, and 72 per cent of primary teachers reported that they felt confident about using IT. The survey indicates that the

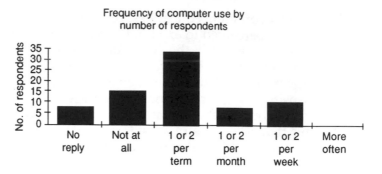

Figure 13.1 indicates one of the reasons for low usage

Figure 13.1   Frequency with which computers are used by geography teachers (GA/NCET, 1993)

teaching style adopted by primary teachers when using computers in the classroom may be important in explaining the higher usage rate.

These results gain some support from a survey undertaken by the Geographical Association in conjunction with the National Council for Education Technology (GA/NCET, 1993). This was a relatively small survey with a low response rate (24 per cent response rate/ 75 individual responses). The relatively low frequency of use is confirmed by this survey (Figure 13.1) which shows that less than 25 per cent of teachers are using computers more than twice a term. However, 12 per cent claim to be using computers once or twice a week. This survey tends to support the idea that there are a few teachers who enthusiastically use IT while the majority are rather more wary. Overall the results highlight the great diversity of IT opportunities for pupils in geography classrooms at the present time.

Figure 13.2 indicates one of the reasons for low usage levels by geographers which persists as a problem in secondary schools. Nearly half of the respondents felt there was limited access to computers in their schools and only 20 per cent had direct control over computers themselves. This suggests that the hardware issue which Shepherd et al. (1980) identified is still a key inhibitor in the mid-1990s.

Another factor for many teachers continues to be the lack of suitable software. This relates to the very high standard geography teachers expect of their resources. It seems that perceptions have changed remarkably little since 1983 when Shepherd referred to teachers who felt that much of the

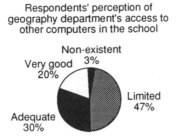

Respondents' perception of
geography department's access to
other computers in the school

Figure 13.2 Access by geography departments to school computers (GA/NCET, 1993)

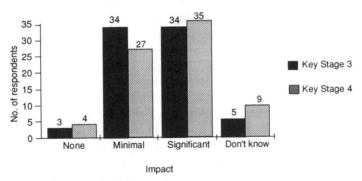

Figure 13.3 Impact of National Curriculum IT Orders on geography teaching (GA/NCET, 1993)

available geography software was disappointing, uninteresting and limited (Shepherd, 1983, p. 52).

Finally, one of the most revealing questions in this survey asked teachers whether they felt that the inclusion of IT as a requirement across National Curriculum subjects was having an impact on their geography teaching. Figure 13.3 shows the surprisingly large number of teachers who felt minimally affected by the IT requirements.

The overall picture presented by these statistical surveys is of a very patchy use of IT in geography lessons. For many pupils the opportunity to use IT is rare and is subject to chance. Despite the existence of a National Curriculum in schools it would be difficult to predict with confidence the IT experience in geography of a school-leaver in 1994. This is a very current issue for many teachers in institutions of further and higher education who are increasingly having to design activities for students who arrive with such varied experiences.

Viewpoint three: IT is an entitlement for pupils taking geography courses

There is a view which is widely gaining acceptance that IT is an entitlement which geography pupils should expect to receive. Furthermore, it is stated that there should be a coherent *minimum entitlement* agreed by geographers which ensures that all pupils receive comparable opportunities. This view emerged and was clarified during a conference sponsored by the DfE in March 1993, to advise on the future of IT in school geography (Rawling for DfE, unpublished).

As a result of this conference, the Geography and IT Support Project was set up (funded by the DfE and jointly administered by the GA and NCET). One of its earliest endeavours was to define what is meant by a minimum IT entitlement for pupils at Key Stages 3 and 4. A pamphlet, *Geography: a Pupil's Entitlement to IT* (GA/NCET, 1994) was published and sent to all secondary schools. A similar publication was produced for primary teachers and has been sent to all primary schools in 1995. These pamphlets make it clear that learning in geography is not only enhanced by the use of IT but may be inhibited if IT is not in use. The five entitlement statements (Figure 13.4) focus on the needs of the geography curriculum and make it clear that these needs may, at times, be best served by using IT as a resource.

The entitlement statements are a powerful tool which give geography teachers a focus for planning in the future. It is significant that the idea of a minimum entitlement for geography and IT has emerged from the DfE, but has been supported through the National Curriculum Orders in geography and in IT and in the SCAA support publications. The geography/IT leaflet produced by SCAA uses the same five entitlement statements as the basis for presenting its guidance for teachers (SCAA, 1995).

The range of viewpoints expressed here about introducing IT into the geography curriculum have proved to be remarkably persistent over the last 15 years (Kent, 1983). If the IT capabilities described in the new National Curriculum Orders are to become reality in geography classrooms, then a number of issues need to be addressed with some degree of urgency.

Pupils studying geography are entitled to use IT:	When undertaking these activities in geography:	IT can contribute by making possible:
to enhance their skills of geographical enquiry	collecting, investigating and questioning data from primary (fieldwork) and secondary sources; undertaking a broad enquiry approach to a topic	*the use of* large amounts of data (e.g. data-handling packages, CD-ROM) and data otherwise difficult to obtain (e.g. data-logging); a wide range of IT techniques and approaches (e.g. creating and selecting maps, graphs for a report)
to gain access to a wide range of geographical knowledge and information sources	drawing on appropriate sources to obtain factual information, ideas and stimuli relating to place, physical, human and environmental topics	*access to* new sources about places and environments (e.g. newspaper on CD-ROM); different ways of viewing the world (e.g. remote sensing); moving images, sound, first-hand contact (e.g. CD-ROM, E-mail), instantaneous images and information (e.g. fax, remote sensing)
to deepen their understanding of environmental and spatial relationships	analysing change over time, locational decisions and people/environment interrelationships	*insight into* relationships otherwise inaccessible to pupils (e.g. modelling packages) and monitoring change over time (e.g. logging weather information)
to experience alternative images of people, place and environment	developing awareness and knowledge of other cultures, places and societies and creatively presenting one's own 'sense of place'	*access to* real images, views, and first-hand contact (e.g. E-mail, CD-ROM, video disk), creative ways of mixing sound, text and images (e.g. multimedia, word processing, graphics)
to consider the wider impact of IT on people, place and environment	studying specific examples, e.g. changes in lifestyle, environmental impacts and locational consequences	pupil knowledge and awareness of IT use and applications in work and society

Figure 13.4   IT entitlement for geography pupils (GA/NCET, 1994)

Planning for Pupil Entitlement in Your School

Having read the leaflet, what can you do next to move towards implementing a pupil entitlement for geography and IT. Here are some suggestions:

1. Review

What do you do already?

Refer to the five entitlement statements to review your current practice.

Can you do more?
Consider staff confidence and expertise, access to computers, availability of software, school IT policy, good teaching ideas.

Evaluating Success
Is IT helping pupils
in geography to:

- handle data more confidently?
- discover patterns and trends more easily?
- choose appropriate IT for their enquiries?
- use a wider range of information sources?
- understand relationships between geographical factors more quickly?
- be more aware of other people's views and 'sense of place'
- recognise the importance of new technology as a geographical factor?

2. Plan

What are the opportunities at course level?
Use the entitlement statements and the practical checklist to highlight opportunities in each Key Stage course.
Plan departmental objectives for geography and IT

How could IT be included in schemes of work?
Use the IT practical checklist to help plan IT applications in schemes of work.

4. Create a Geography/IT Action Plan

With:

- clear objectives over a realistic timescale (1 or 2 years)
- recognition of the whole school situation
- reference to Key Stage and year group course plans
- reference to specific schemes of work
- identification of resource and inset provision
- reference to ways of monitoring progress and checking achievement (departmental and pupil)

3. Identify

What else is needed to accomplish this?
Outline the professional development needs of the department. Look at whole school IT policy and assessment needs. Consider the hardware and software requirements.

Who can help?
IT co-ordinator, senior management team, other heads of department. LEA Advisors (IT/geography), IT centre. The Geographical Association and the National Council for Educational Technology.

Figure 13.5   Planning for pupil entitlement in your school

Issues for the immediate future

Geography teachers need help with planning a pupil entitlement for IT which will work across each key stage

It is widely recognised that, even when a teacher or department supports the entitlement viewpoint, further practical guidance will be needed. To help guide teachers a four-stage plan culminating in a geography/IT action plan has been proposed for secondary school geography departments (Figure 13.5).

The curriculum planning approach recognises the centrality of the geography department in the decision-making process, but also emphasises the need to draw on a wide range of support from within and outside the school. A similar planning approach is being suggested for primary schools (GA/NCET, 1995). Whether this would be an appropriate approach in institutions of further and higher education will be discussed later.

Geographers must have access to and information about good teaching materials which incorporate IT activities and mark schemes

Publishers are producing a wide range of material (including assessment items) for the geography National Curriculum, but how often does this contain any reference to IT? Well-designed materials which have geography as the *primary* focus, but which also incorporate ideas on using IT are needed. These may be materials which are incorporated into geography textbooks or may be self-contained packages targeted at particular aspects of the programmes of study.

Two examples of recently published resources illustrate the type of material which is needed. In the first example the IT is an integral part of a geographical enquiry and was designed to be part of the assessment process for a GCSE module.

Example 1   *Farm Studies and Enquiries* (NCET, 1992)

This package was developed in 1988/89 before the publication of the National Curriculum in geography. However, many of the issues explored are relevant to Key Stage 3 geography. The package contains examples of teacher-planned enquiries based on one farm. Pupils can investigate changing land use, role-play a day in the life of the farmer and consider the decision-making processes involved in farming. In one enquiry the pupils evaluate five sites on the farm where the farmer might locate a camp site. The package includes the pupil worksheets, the teacher's mark scheme and the pupil evaluation forms. Detailed advice is given on how to use the software. The IT used includes databases and modelling covering two strands of the National Curriculum. Significantly, ideas about assessing IT are built into the geography assessment in the enquiry.

In the second example the activities were purposely designed to give geography a role in delivering and assessing the National Curriculum for IT. One aspect of integrating IT across key stages which is seldom addressed in published material is the need to ensure progression. In this package, ideas on progression in geography and in IT are presented together.

Example 2   *The WATER/EXCEL Project* (NCET, 1994)

Two packages have been published. The first is for Key Stage 2 and focuses on the use of water in the home and on what happens to water when it reaches the ground. The second is for Key Stages 3 and 4 and focuses on flooding and on the dynamics of two small streams.

These packages were designed and tested after the original IT and geography Orders had been published. The materials refer very precisely to the old statements of attainment and to the programmes of study in both documents, but they still have direct relevance to the new Orders. They use one widely available piece of software (Microsoft EXCEL) and explore its potential for data handling, date presentation and modelling at each key stage. There are schemes of work, pupil materials and assessment tasks. Records of achievement for pupils and recording devices for teachers are included.

In planning for IT use, geographers need special help with teaching and learning approaches and classroom organisation

Many secondary teachers complain of a lack of support when attempting to introduce IT into their lessons. There is no doubt that teaching approaches and classroom organisation need to be rethought on many occasions when IT is in use. For example, for many, a room full of computers allowing every pupil to undertake the same activity at the same time is seen as desirable, if not essential. Strategies for time sharing with one to two computers in a class are often ignored. However, it is possible to use computers in this way as primary school pupils illustrate every day.

As an example of how this approach might be extended to secondary schools it is useful to consider the use of CD-ROM in the classroom. Very few teachers have access to more than one CD-ROM player at a time and sharing is therefore essential.

Environmental Pressures in a National Park

A year 7 group is being introduced to a range of basic map skills and the teacher decides to do this by designing an enquiry investigating environmental pressures on localities in the Lake District. The pupils are set a range of activities to complete over two weeks. To utilise the Langdale CD-ROM (Creative Curriculum Software, 1994) on one computer in the classroom, each pair of pupils has one task to complete on the computer and five to complete using other maps and photographs available in the classroom. Mapwork skills are built into the exercises provided by the publishers on worksheets. The worksheets are provided as files on disk, allowing teachers to edit them to suit the needs of particular ability groups. For example, the files may be edited so pupils will simply have to fill in missing words in the prepared text. Differentiation is made possible by encouraging some pupils to use the original questions and others to use the incomplete sentences.

In order to 'track' the use of the computer by individual pupils and to monitor the development of IT competence there is a log kept by the computer and filled in by each user at the end of their session.

This example illustrates the flexibility that is possible when using IT resources. However, the organisation of the classroom activity and the production of resources is time consuming and the use of in-service training opportunities to develop this type of material is vital. The current average of less than one day in two years given to IT training for geographers in schools (GA/NCET, 1993) is clearly not sufficient.

Geographers need to consider the advantages and the pitfalls in becoming central players in the delivery of IT and the assessment of IT in the school

Many schools now have considerable resources devoted to IT and geographers must decide whether they wish to be identified as priority users of IT with access to these resources. Central to this decision is the need to be committed to assessing part of each pupil's IT capability. When geographers become involved in assessing IT capability then some crucial questions arise:

1. Which aspects of the IT National Curriculum can most appropriately be assessed through geography?
2. What should the balance be between assessing *geography through the use of IT* and assessing *IT through geography?*
3. Can geographers realistically manage both at once?

The result of these deliberations has been a wide range of different experiences for pupils. Figure 13.6 attempts to indicate what has been happening in schools. Most geographers will be able to place themselves somewhere along the continuum indicated. More importantly, perhaps, geography departments can use the table to consider where they would like to be in a few years' time.

Nationally it is becoming clear that geography is seen as a subject which can successfully be involved in delivering and assessing IT. As a result, funding has been made available (through the GEST budgets) for in-service training to enhance the role played by geography in developing IT. The threat of declining timetable allocations, which may result from geography's optional status in the National Curriculum at Key Stage 4, may encourage geographers to make the use of IT a high priority. This could be one way of ensuring the retention of the subject on the timetable.

	Stage One	Stage Two	Stage Three	Stage Four
Geographers provide	the minimum statutory contribution required by the Geography National Curriculum	an enhanced contribution within the Geography National Curriculum	IT experiences to reinforce and support IT capability in the National Curriculum	part of the statutory requirement for IT capability in the National Curriculum
Using IT	• Very little use of IT • Lack of continuity and progression • Conforms to requirements of programme of study though very infrequent use of IT in enquiries	• IT is an integral part of the geography schemes of work • Range of generic and subject-specific resources used • Balance and progression within geography are ensured by careful design of work units in each year	• IT is mapped into geography to support other providers • Main intention of geographers is to reinforce pupils' experiences in other subjects	• Geography is an integral part of the IT scheme of work • Geographers identify slots in schemes of work in which IT capability is an objective • There may sometimes be a need to 'divert' from the essential elements of the Geography National Curriculum in order to meet these objectives
Assessing IT	• No assessment of IT made as statements of attainment in original Geography Order did not require this	• Informal monitoring of class use of IT • IT capability is incidental to main objectives • Work is marked and records kept of progress in geography but not in IT	• IT co-ordinator is consulted in the final planning stages of geography units • IT capability may be brought into sharper focus as a result of consultation • Geographers do not contribute *evidence* of IT capability but reinforce the work of other teachers who are collecting evidence	• Evidence is gathered by geographers for assessment of IT capability • IT co-ordinators rely on evidence from geographers (and others) for—making judgements about pupil performance in IT at the end of the key stage

Figure 13.6 Using and assessing IT through geography at Key Stage 3: a range of current positions. *Source:* Burkill, 1993 (revised version of a paper presented at a conference (1993) organised by the Department of Education to discuss the future of IT in geography)

The implications of an IT entitlement in school geography for higher education

If it is reasonable to assume that the issues raised in the previous section are going to be effectively addressed and that geography will play an increasing role in developing IT, then the capability of geography students entering higher education will be substantially increased. Students will have greater confidence as well as competence in using both subject-specific and generic software. They will also be used to greater flexibility in the way they work and will expect more control over their own learning. The need to maintain the momentum provided by the school curriculum into and through higher education will become a major issue in the next five years. The key question raised below is one that needs to be addressed by geography departments in higher education in the near future.

Will pupils who have had an entitlement to IT in their school geography courses be expecting an entitlement in higher education? How should universities and colleges respond?

If higher education does decide to address this question there will be a number of issues to consider:

1. As the 'exit competences' of many school pupils become more sophisticated, it will be a major challenge to decide how to build on this. For example, many pupils moving into higher education are becoming familiar with the hardware (PCs, Apple computers) and the software (Windows, Windows applications such as Microsoft EXCEL) which are the backbone of undergraduate courses. How should those institutions which provide basic IT awareness courses react? Are these courses redundant or at best optional?
2. At present the range of IT experiences offered by school geography departments to pupils has probably never been greater. This suggests that HEIs will be faced with mixed experience and mixed ability groups for some years to come. How should they cater for this disparate range of experiences? Can experiences of successful teaching and learning strategies be shared between institutions?
3. Pupils who have used IT in school geographical enquiries

will be used to classrooms where teachers do not dominate activities and they will be used to situations where learning is divergent and often in their own control. Does higher education have the teaching and learning strategies in place to take on this generation of independent learners? Will the software being developed for use in geography departments in higher education (for instance by the Teaching and Learning Technology Project at the University of Leicester) prove to be appropriate for independent learning?

These questions are not easily answered and have perhaps been left unaddressed too long. The concept of a minimum entitlement for IT in geography which has been fostered in the school sector is one to which higher education should give serious consideration. Many will argue that there is too much diversity in higher education to obtain a single response, but schools are very diverse institutions as well. Given the increasing trend to accountability and quality control in higher education, it may be that the definition of an IT entitlement for higher education geography would be a sensible pre-emptive strategy, ensuring the contribution and well-being of geography on the wider stage of tertiary education.

References

Creative Curriculum Software (1994) *Langdale—CD Rom*. Creative Curriculum Software, Halifax.

DES/WO (Department of Education and Science/Welsh Office) (1990) *Technology in the National Curriculum*. HMSO, London and Cardiff.

DES (1991) *Geography in the National Curriculum*. HMSO, London.

DfE (Department for Education) (1993) *Statistical Bulletin: Survey of IT in Schools*. March. DfE, London.

DfE (1995a) *Geography in the National Curriculum (England)*. HMSO, London.

DfE (1995b) *Technology in the National Curriculum*. HMSO, London.

DfE (1995c) *IT in Schools: Statistical Survey 1994*. DfE, London.

GA/NCET (1993) *A Survey of IT in Geography*. NCET, Coventry.

GA/NCET (1994) *Geography: a Pupils's Entitlement to IT*. NCET, Coventry.

GA/NCET (1995) *Primary Geography: a Pupil's Entitlement to IT*. NCET, Coventry.

Kent, W.A. (ed.) (1983) *Geography Teaching and the Micro*. Longman, Harlow.

Kent, W.A. (1992) The new technology and geographical education. In M. Naish (ed.), *Geography and Education*, Institute of Education, University of London.

NCET (National Council for Educational Technology) (1990) *Learning Geography with Computers* (Pack) (2nd edn). NCET, Coventry.

NCET/HIT Project (1992) *Farm Studies and Enquiries*. Longman, Harlow.

NCET (1994) *The Water Excel Project Key Stage 2, 3 and 4*. NCET, Coventry.

Rawling, E. (1993 unpublished) *Geography and Information Technology in Schools*. Report of a conference organised by the DfE held in Birmingham, Mach 10–12, 1993.

SCAA (Schools Curriculum and Assessment Authority) (1995) *Information Technology in the National Curriculum. The New Requirements: Key Stage 3* and *Information Technology in the National Curriculum. The New Requirements: Key Stages 1 and 2*. SCAA, London.

SEAC (Schools Examination and Assessment Council) (1993) *Pupil's Work Assessed: Information Technology Capability at Key Stage 3*. SEAC, London.

Shepherd, I.D.H. (1983) The agony and the ecstacy – reflections on the microcomputer and geography teaching. In W.A. Kent (ed.), *Geography Teaching and the Micro*. Longman, Harlow, pp. 51–64.

Shepherd, I.D.H., Cooper, Z.A. and Walker, D.R.F. (1980) *Computer Assisted Learning in Geography*. CET/GA, London.

Watson, D. (ed.) (1994) *The ImpacT survey: An Evaluation of the Impact of Information Technology on Children's Achievements in Primary and Secondary Schools*. DfE/King's College, London.

WO (Welsh Office) (1991) *Geography in the National Curriculum, Wales*. HMSO, Cardiff.

WO (1995) *Geography in the National Curriculum, England*. HMSO, Cardiff.

Geography in the Scottish school curriculum

14

Leslie Hunter

Curriculum change is going on apace in Scotland and much of this mirrors developments in England and Wales, but with some significant differences. To begin with, the National Curriculum, GCSE, A and AS levels do not apply to Scotland. The main purpose of this chapter is to outline developments in Scotland, not to compare them with developments in England and Wales other than for clarification.

Curriculum and assessment 5–14

The National Curriculum in England and Wales is bound by Statutory Orders and therefore is mandatory. The parallel development in Scotland is a set of national guidelines on curriculum and assessment covering all aspects of education in the age range 5–14. This therefore includes seven years of primary school and the first two years of secondary school. These are guidelines which are not mandatory. All 12 of the Scottish local education authorities (LEAs) have agreed to implement them, but at their own pace within an overall timetable suggested by Her Majesty's Inspectors of Schools (Scotland) which stretches to 1999.

Guidelines have been produced for five major curriculum areas: English language, mathematics, expressive arts,

Geography into the Twenty-first Century, Edited by E.M. Rawling and R.A. Daugherty,
© 1996 John Wiley & Sons Ltd.

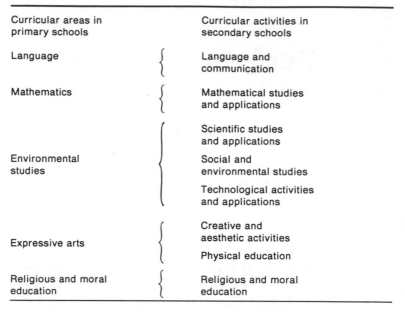

Curricular areas in primary schools	Curricular activities in secondary schools
Language	Language and communication
Mathematics	Mathematical studies and applications
Environmental studies	Scientific studies and applications
	Social and environmental studies
	Technological activities and applications
Expressive arts	Creative and aesthetic activities
	Physical education
Religious and moral education	Religious and moral education

Figure 14.1 Progression to secondary schools. *Source:* 5–14 Environmental Studies Guidelines

religious and moral education and environmental studies. For each of these areas a set of attainment outcomes has been provided and minimum time allocations recommended, with a flexibility factor of 25 per cent. This flexibility factor allows schools to adjust the balance of their curriculum to best meet the needs of their pupils within the guidelines. Geography and history do not appear as separate subjects within the guidelines but geographical, historical, sociological, political, scientific, technological, health, information technology (IT), home economics and environmental education components are included in environmental studies which is allocated 25 per cent of curriculum time in primary and up to 30 per cent of curriculum time in secondary schools. The relationship between these five areas and secondary subject provision is shown in Figure 14.1.

In all schools the priority has been to implement the guidelines for English language and mathematics and the rolling programme for development of the other three curriculum areas has been left to individual schools or groups of schools. The last of the guidelines to be published in March 1993 were for environmental studies. For the first time a set of attainment outcomes has been provided for this complex area of the curriculum which up until now had been covered in a rather haphazard way by many schools. The guidelines require close co-operation over the Primary 7 (P7)–Secondary 2 (S2) stages and also between the subject departments of the secondary school. The substantial curriculum change needed to implement the environmental studies guidelines is getting under way.

The environmental studies programme is in the form of a set of attainment outcomes and targets within the five components of science, social subjects, technology, health education and IT. For each of these a set of attainment outcomes with key features has been developed. Within the social subjects area of the curriculum, the attainment outcomes are: 'Understanding people and place', 'Understanding people in the past' and 'Understanding people in society'. The key features for environmental studies are stated in contexts and content at three broad stages: P1 to P3, P4 to P6 and P7 to S2, thus clearly linking the continuity, coherence and progression of the curriculum development across the upper primary and lower secondary stages. Each attainment outcome is further set out in strands: knowledge and understanding, planning, collecting evidence, recording and presenting, interpreting and evaluating and developing informed attitudes.

The geographical component is mainly to be found in 'Understanding people and place', but a significant part of physical and systematic geography is in the science attainment outcome 'Understanding earth and space'. The structure of these attainment outcomes is shown in detail in the guidelines for each key feature. An example of this is the key feature of 'aspects of the physical and built environment' in 'Understanding people and place' which is set out in terms of contexts and content at the P1 to P3, P4 to P6 and P7 to S2 stages.

In many ways the recognition of the scientific element of geography is to be welcomed, but of course, it is only when the interrelationships between physical and human aspects of geography are explored and developed that the holistic view of the subject is attained. It is worth reflecting that the age range is 5–14 and that the way in which the curriculum is delivered differs in primary and secondary schools. In general, one teacher will cover all or at least most of the aspects of environmental studies in primary school, whereas separate subject departments prevail in secondary schools.

As a result, in the secondary sector subject departments of science and geography will have to agree how to teach and who teaches those aspects of physical geography in the science outcomes which have been traditionally taught in, and are essential to, geography. This need not be an issue, and indeed in many schools good co-operation already exists between geography and science departments. What is important is to ensure that pupils receive quality learning experiences in this area of the curriculum; this is the key factor, not demarcation.

The introduction of the 5–14 programmes in English language and mathematics is well under way in most primary schools and are in place or advancing rapidly in many secondary schools. With considerable local and national support teachers have responded successfully to the 'attainment outcome' approach and this should stand them in good stead as the other areas of this development are progressively introduced.

Provision in the 14–16 age range

In the middle stages of secondary school (S3, S4) there have been major changes in the curriculum unique to Scotland. In the 1980s over a period of years Ordinary (O) grade courses (equivalent to O level) were phased out and Standard (S) grade introduced. The social subjects, including geography, came on stream in 1990.

O grade was originally aimed at the top 30 per cent of the ability range; this was gradually extended to the point where some 70 per cent of pupils took a course and sat an examination designed for the top 30 per cent. Clearly, this was unsatisfactory. O grade was largely norm-referenced assessment by external examination with an 'appeals' system. There was no examined practical or investigative element and no examination of the understanding of processes in geography such as landscape evolution. Despite a major overhaul of the syllabus in the early 1970s, O grade geography was fundamentally unchanged; it relied heavily on recall and did not examine processes in geography.

S grade geography was introduced as a unitary syllabus at three levels, Foundation (F), General (G) and Credit (C), to provide meaningful courses and certification for all levels of ability. Assessment is by a set of grade-related criteria over

- 7 fundamental concepts, e.g. location, interdependence
- 17 key ideas grouped in 3 study themes
- emphasis on enquiry-based learning
- 3 levels of certification: Foundation, General, Credit
- 3 assessable elements: knowledge and understanding; evaluating; investigating
- internal and external assessment

Examples

Study theme	The physical environment
Key idea 1	Physical landscapes are the product of natural processes and are always changing
Study theme	The human environment
Key idea 8	Urban settlements have dynamic patterns relating to their size, form and function
Study theme	International issues
Key idea 17	Schemes of self-help, along with national and international aid, seek to encourage social and economic development

Note: In external examinations questions may draw on more than one key idea across the 3 study themes

Figure 14.2   Standard grade

three elements of knowledge and understanding, evaluating and investigating. It includes both internal school-based assessment and external examination with an equal weighting for knowledge and understanding and evaluating; internal assessment of investigating is externally moderated by the Scottish Examination Board (SEB). The appeals system has been retained. The syllabus was built around 21 key ideas grouped into three study themes of 'The physical environment', 'The human environment' and 'International issues'. The key ideas were designed to allow the exploration of seven concepts central to geography, but these concepts are not explicitly assessed for certification. External examination questions can draw on more than one key idea and one study theme (see Figure 14.2).

The syllabus sets out a clear and comprehensive set of aims for geography and much of it was innovative. There was, for the first time, a compulsory investigation for all levels of ability, processes in geography were progressively examined from F, to G, to C and environmental issues and environmental education were given a higher profile. But underlying the whole S grade programme was the need for pupils of all levels of ability to be more involved in their own education. It was largely to do with 'learning and teaching', particularly to engage pupils in an enquiry-based approach to learning.

Inevitably with such a radical change of syllabus there were problems, despite considerable government and LEA support in the provision of resources and staff development. There was a general feeling among geography teachers of 'content overload' and as a result the number of key ideas was reduced to 17 with some conflation and some excision of content. The investigation was also a cause for concern to some teachers and as a result steps have been taken to clarify the requirements of assessment and to produce graded sample investigations. One great advantage of a national examination board (SEB) is that proposed changes can be put out for consultation and fairly quickly implemented.

The impact of S grade has been great. There has been a marked move towards more enquiry-based learning and to investigative work, not only in S3 and S4 but also in S1 and S2 and many geography departments have introduced an assessment system in S1 and S2 based on the S grade assessable elements. Now almost all pupils sit certificate examinations. It is of interest to note that geography has for many years been the most popular social subject at S3 and S4 as the table shows (see Figure 14.3). More boys than girls opt to study the subject, an imbalance that requires further research. S grade is now firmly established and is to remain the main form of provision underpinning the 'Higher Still' proposals outlined below.

Early in 1995 a report entitled *Effective Learning and Teaching in Geography* was published by HMSO as part of a series of subject reports (Scottish Office Education Department, 1995). This is based on HMI reports of some 150 departments over the last 10 years and provides clear evidence of the changes that have taken place in the teaching of geography.

Provision in the 16+ age range

Coming hard on the heels, and directly linked to S grade, revised syllabuses at Higher (H) grade and the Certificate of Sixth Year Studies (CSYS) were introduced in 1991 and 1992 respectively. A primary aim was to build in progression from S3 to S6. These syllabuses had to be acceptable to the Scottish Universities Council on Entrance (SUCE), especially the H grade one as it is recognised as an entry qualification.

The revised H grade syllabus consists of compulsory physical and human cores, a structured choice of three from

Presentations in the social subjects		
Presentations 1993	S grade	Higher in S5
History	18 692	4335
Modern Studies	12 187	2848
Geography	22 278	4968

Figure 14.3 Revised Higher grade presentations in the social subjects

Revised Higher grade geography	
Physical core (systems approach)	• Atmosphere • Hydrosphere • Lithosphere • Biosphere
Human core	• Population geography • Rural geography • Industrial geography • Urban geography
Applications Group 1	• Rural land resources • Rural land degradation • River basin management
Group 2	• Urban change and its management • European regional inequalities • Development and health
Three are studied for examination, one from each of Group 1 and Group 2 and one other.	
The investigation	• Linked to 7 concepts of S grade • Library and/or fieldwork • Product in the form of a report
Assessment External Exam Paper 1 Paper 2 Internal and external moderation	• Physical and human cores: 35% • Applications: 45% The investigation: 20%

Figure 14.4 Revised Higher grade geography; syllabus structure

six applications of geography and an investigation (see Figure 14.4). While the syllabus was widely approved by teachers there has been concern about overload. As a result the SEB consulted schools and other interested bodies about the retention of the investigation and it is to be dropped in and after 1996. This is of particular interest in the light of the wide-reaching changes proposed in post-16 education in Scotland, considered below. The changes to CSYS were less far reaching and largely retained the tripartite structure of a field study, dissertation/folio of essays and a written examination on a specified area (at present China or the USA and Canada) with five prescribed study themes assessed externally in the ratio 3:4:3. Assessment at H grade is by an external examination and an internally assessed and externally moderated investigation. The assessment of CSYS is wholly external.

National Certificate modules

An ever-increasing number of young people are staying on at school beyond the minimum leaving age of 16, not all of them interested in, or capable of following, H grade courses. They have to be provided with alternatives and geography is no exception to other subjects. Through the Scottish Vocational Education Council (SCOTVEC) a large number of National Certificate (NC) modules has been drawn up. These modules have a set of learning outcomes which are internally assessed with external verification by SCOTVEC and are designed to be taught in 40 hours (some half and some double modules have also been drawn up, but the majority are 40 hours). While modules were initially designed to be taught in colleges of further education the possibility of them being taught in schools was also taken into account. Most of the modules were vocationally orientated, but many had a considerable amount of geography in them and were of interest to schoolteachers, for example modules on travel and tourism, and an increasing number of schools made use of them. As a result of demand by schools for relevant modules with more specific geography content a series 'People and the environment' was drawn up. This comprises of a set of 40-hour modules at different levels of difficulty and designed to provide for the study of geography at local to world scales and investigative work. These modules are becoming increasingly popular in schools and serve several purposes, including a preparation of pupils studying them in S5 who may progress to study H grade in S6.

Higher still

In March 1994, the government's response to the Howie inquiry into post-16 education in Scotland was published, entitled *Higher Still, Opportunity for All (SOED, 1993a)*. This set in motion a major review of post-S grade provision and inevitably has a knock-on effect on provision lower down the school. A major feature of the proposals is that courses for all levels of ability have to be provided in modular form for all subjects. The main thrust of the 'Higher Still' programme is to provide appropriately challenging courses for pupils of all abilities and to better prepare students for entry to further and higher education. This will be attained by merging the SCOTVEC modular system with that of the SEB. All courses will be written in modular form with both internal and external assessment. As part of these proposals, courses will be offered at five levels to provide a progression from courses taken in S4. The academically more able pupils will proceed from credit level at S grade in S4 to either a one-year H grade course or a two-year Advanced Higher course. An important part of the 'Higher Still' proposals is that courses will largely be based on existing SEB courses and SCOTVEC modules. These courses will be equivalent to the present S grade, H grade and CSYS levels of attainment and will be the basis for entry to further and higher education.

This development is at a very early stage of development, but it will have major implications for the restructuring of geography provision in the upper secondary school. Within these changes close attention will be required to ensure that continuity, coherence and progression in geographical education are sustained.

Conclusion

Over the past 10 years or so major changes have taken place and are continuing (see Figure 14.5). At the 5–14 stages, geography does not appear by name as a separate subject, but the building blocks are there and it is up to teachers of geography to ensure that the geographical component of pupils' education is effectively delivered. There is now a set of guidelines which can provide continuity, progression and coherence. The progression from 5–14 to S grade requires careful development. Up until now geography has been the most popular social subject at S grade (and all pupils must study at least one social subject). This has been largely due to stimulating courses and good teaching. Progression from S grade to revised H grade and CSYS has been an integral part of planning syllabuses, as have 'balance' and building in

5–14 Environmental Studies Guidelines on Curriculum and Assessment (March 1993 to be in place 1999?)	Geography located in 2 main attainment outcomes: People and place Understanding earth and space
Standard grade Replaced O grade (1990–92)	Unitary syllabus for all levels of ability based on 17 key ideas and 7 concepts Grade-related criteria Assessment by three elements: Knowledge and understanding Evaluating Investigating
Revised H grade Replaced Alternative Higher (1991–94)	Core consists of physical, human and applications Assessment by exam and investigation
Revised CSYS Relatively little change (1992–94)	Tripartite structure: Field study Dissertation/folio of essays Written exam on specified areas
SCOTVEC modules Increasing range (1988 onwards)	Increasing number with geographical content, e.g. Series 'People in the Environment'
Higher Still Government response to 'Howie Report' (March 1994)	Major changes in the 16+ provision Moves towards modular courses over 1 or 2 years

Figure 14.5   Summary of recent developments in Scottish education

'up-to-date' approaches to geography. It is important that these are retained in whatever courses develop as a result of the 'Higher Still' programme.

Geography is and must be a dynamic subject prepared to take a hard look at itself and be ready to implement change to meet the needs of a changing society. It was O.E. Baker, in describing developments in the crop belts of the USA in the 1920s who said, 'there is nothing more permanent than change'. That certainly applies to curriculum development in Scotland. Both the subject of geography and teachers of geography are robust enough to respond positively to the challenging demands of change. Teachers will continue to provide interest, relevant and enjoyable learning experiences in geography for young people.

Bibliography

HMI (Her Majesty's Inspectorate) (Scotland) (1994a) *5–14: A practical guide for teachers in primary and secondary schools.* Scottish Office Education Department, Edinburgh.

HMI (Scotland) (1994b) *Implementing 5–14: a Progress Report by HM Inspectors of Schools.* Scottish Office Education Department, Edinburgh.

SEB (Scottish Examination Board) (1988) *Standard Grade Arrangements in Geography.* SEB, Dalkeith.

SEB (1990) *Scottish Certificate of Education Higher Grade and Certificate of Sixth Year Studies, Revised Arrangements in Geography.* SEB, Dalkeith.

SOED (Scottish Office Education Department) (1993a) *Higher Still: Opportunity for All.* HMSO, Edinburgh.

SOED (1993b) *Curriculum and Assessment in Scotland: National Guidelines, Environmental Studies.* HMSO, Edinburgh.

SOED (1995) *Effective Learning and Teaching in Scottish Secondary Schools, Geography.* HMSO, Edinburgh.

Geography 5–19: some issues for debate

15

Eleanor Rawling

Introduction

In January 1995, the process of revision to the National Curriculum in England and Wales, begun by Sir Ron Dearing in 1993, officially came to an end with the publication of the revised Statutory Orders for all 10 National Curriculum subject (11 in Wales). If government rhetoric can be believed, this situation will remain stable for the next five years. After nearly eight years of turmoil and change, 5–14 geography may be entering a period of greater calm, or, at least, teachers may be provided with the opportunity for some professional reflection and rethinking rather than merely responding to centrally imposed debate and direction.

The situation is not so settled for 14–19 geography. The decision in July 1994 to make geography an optional subject at 14–16 years means that the subject must compete against numerous claims on curriculum time at Key Stage 4, including those of newly established vocational courses. The apparent stability of A/AS level geography courses and indeed, of the whole 14–19 curriculum, was also thrown into doubt by the annoucement (April 1995) that Sir Ron Dearing would lead a review of 16–19 qualifications during 1995–96. Although a new set of A/AS syllabuses has been approved by the School Curriculum and Assessment Authority (SCAA), geography, along with other subjects, must expect more change in the medium and long term.

Geography into the Twenty-first Century, Edited by E.M. Rawling and R.A. Daugherty,
© 1996 John Wiley & Sons Ltd.

This chapter will demonstrate that, for the first time, there are central guidelines for geography throughout the 5–19 age range (National Curriculum, GCSE Criteria, A/AS Subject Core) although they form only a minimum framework and each is different in format. Analysis of these guidelines provides an opportunity to reflect on the broad character of geographical content and approach established by central diktat, and to draw out some key issues for schools and higher education. The questions to be addressed in this chapter are:

1. What are the key features of geography for the 5–19 age group?
2. How are the character and status of school geography affected by the wider context of curriculum change?
3. What are the assumptions which lie behind the sequence of content 5–19 and why are these significant for higher education?
4. Is there a need for a greater shared understanding about the study of place and the development of geographical enquiry, in order to ensure quality and continuity in the curriculum?
5. How far does school geography contribute to the development of broader skills and abilities? Does the situation change in higher education?
6. Do the central guidelines for geography 5–19 provide an acceptable balance between central control and professional freedom? What are the implications for higher education?

Each question is developed under the headings which follow.

Given the curriculum upheaval of the early 1990s and the likelihood of further change heralded by the 'Second Dearing Review', it is even more pressing that schools and higher education engage in an informed debate about these issues.

Key features of geography in the 5–19 curriculum

For the first time, there are now centrally provided guidelines for school geography in the 5–19 age range and Figure 15.1 provides a summary of their content and an interpretation of their character. For 5–14 pupils, geography is a compulsory subject in the National Curriculum and all maintained schools in England and Wales are statutorily bound by the requirements set out in the Orders for Geography (DfE, 1995; WO, 1995). The content to be taught is specified in *programmes of study* for each of three age groups; Key Stage 1/5–7 years, Key Stage 2/7–11 years, Key Stage 3/11–14 years. The programmes of study contain information about the places, themes and skills to be studied, but allow teachers considerable freedom in planning the details of the geography curriculum.

At Key Stage 1 (5–7 years) pupils are required to study two localities, one of which must be the local area. In England, they also study environmental quality, although a choice from three themes, including environmental quality, is available in Wales. Key Stage 2 pupils (7–11 years) are required to study three places at locality scale and four aspects of thematic study focusing on rivers, weather, settlement (or economic activities in Wales) and environmental change. Finally at Key Stage 3 (11–14 years) pupils must study two countries in different states of development (and also Wales is specified in the Welsh Order) and a range of themes which cover many aspects of physical and human geography including tectonic processes, weather and climate, economic activities and environmental issues. Certain features are common to all key stages and all of these are inherited from emphases in the original 1991 Order. Geographical work must cover a balance of place and thematic study; an enquiry approach to geographical work should be employed; pupils should be introduced to a range of geographical skills and techniques including mapwork, fieldwork, use of secondary sources and IT; pupils should be provided with a basic framework of locational knowledge; and opportunities should always be taken to develop awareness of the wider context (local, regional, national, international, global) within which people and places operate.

The other key element of the revised National Geography Curriculum in England and Wales is that a single *attainment target* (geography) defined by eight level descriptions, plus one description of exceptional performance, draws attention to pupil performance in the key aspects of knowledge, understanding and skill development in geography. The level descriptions provide the basis for teachers to make judgements at the end of a key stage about the work being done by individual pupils. At the time of writing, statutory

	Key stage 1/5–7 years (NC)	Key stage 2/7–11 years (NC)
Character and purpose of geography	— building on own experience and skills to reach better understanding of own local surroundings and some knowledge and awareness of the wider world	— widening the range of scales and contrasting environments studied, to gain deeper understanding of places and to investigate aspects of physical and human geography
	[understanding my own place and becoming aware of the wider world]	*[growing awareness of how my place compares with other places and how geography helps me understand the world around me]*
Geographical enquiry and investigative work	**focus on** asking: — what/where is it? — what is it like? — how did it get like this? opportunities to: — observe, question, record — communicate ideas and information	**focus on** asking: — what/where is it? — what is it like? — how did it get like this? — how/why is it changing? opportunities to: — observe, ask questions — collect and record evidence — analyse evidence, draw conclusions, communicate findings
Prescribed content	**Places** — locality of school and a contrasting locality **Theme:** — *in England*, environmental quality — *in Wales* either weather or jobs/journeys or environmental quality	**Places** — locality of school, a contrasting locality in the UK, and a locality from a country in a different state of development **Themes:** — rivers — weather — environmental change — settlement (*England only*) — economic activities (*Wales only*)

Awareness of wider world and context of places/themes studied

Figure 15.1 Geography in the 5–19 curriculum: a summary of its character and content. (Information derived and interpreted from National Curriculum Orders, GCSE Criteria and A/AS Subject Core)

Key stage 3.11–14 years (NC)	14–16 years (GCSE)	16–19 years (A/AS)
— broadening and deepening geographical understanding and skills, through study of a wide range of places and themes at a full range of scales. Increasing ability to recognise processes, patterns and trends in geography	— growing ability to understand the character of places and environments, to be aware of geography's contribution to understanding issues, and to form own views on environmental and spatial matters	— deeper understanding of particular aspects of geography, and growing competence in applying wide range of knowledge, skills and understanding to real world issues and problems — greater stress on global issues, interrelationships and interdependence
[greater awareness of the value of geography in explaining features and places, and in giving a wider national and global context]	*[using geography to make sense of the world and to form views and opinions based on evidence and understanding]*	*[competence as a geographer and ability to use and apply this to real world]*
focus on asking: — what/where is it? what is it like? — how did it get like this? — how/why is it changing? — what are the implications? opportunities to: — identify geographical questions — establish appropriate sequence of investigation — identify evidence and collect, record and present it — analyse and evaluate — draw conclusions and communicate findings	**focus on:** a range of geographical enquiry skills including: — identification of questions/issues — establishing appropriate sequence of enquiry — identifying and collecting evidence — recording and presenting it — describing, analysing, interpreting evidence — drawing conclusions and communicating findings — evaluating method and evaluating conclusions	**focus on:** ability to produce a complete enquiry, including abilities to: — collect, record, interpret evidence — select from enquiry methodologies — organise, present, communicate information — analyse, synthesise, critically evaluate — use quantitative and qualitative investigating techniques
Places: — two countries in different state of development other than UK plus *in Wales*—Wales **Themes**: (full range of scales) — Tectonic processes, geomorphological processes, weather and climate, ecosystems, population, settlement, economic activities plus *in England* Development, Environmental issues *in Wales:* Management of environment Resources and environment Global environment	**Syllabuses must require:** — balance of physical, human and environmental aspects — range of scales (local–global) environments, and different states of development — physical and human processes — people/environment interrelationships — geographical aspects of issues — significance of attitudes/values in decision-making — locational knowledge and terminology — range of geographical enquiry skills	**Syllabuses must require:** — a theme which emphasises interaction between people and environment — a chosen physical environment — a chosen human environment — personal investigative work (based on first-hand ('A' only) and second-hand data) All this at full range of scales and drawing on a variety of different places and contexts for study

teacher assessment using the level descriptions will (subject to consultation) be required in 1997 for 14-year-old pupils (end of Key Stage 3), but no decisions have been made about whether national tests will be reintroduced for geography. Given the difficulties in reaching satisfactory and practicable arrangements for testing the core subjects during 1994 and 1995, it seems unlikely that national tests will be reintroduced for the non-core subjects, at least in the foreseeable future. At Key Stages 1 and 2, there is no requirement for teacher assessment or national tests either in 1996 or 1997 and decisions about future years remain to be made.

Because the revised National Curriculum has been planned as a coherent 5–14 programme, certain elements of progression are apparent. They are referred to in the introductory paragraphs to each programme of study, and they are implicit in the required content and in the make-up of level descriptions. Effectively, progress in geographical work 5–14 is related to an increase in scales of study, a growing range and complexity of topics, places and environments studied, increasing attention given to process in geography and growing competence and independence in geographical enquiry. Although breadth and depth of conceptual under-standing are hinted at, the development of key ideas in geography is left to the teacher. In fact, probably the most significant feature of the 1995 Orders is that, unlike the 1991 Orders, they genuinely comprise a minimum national entitlement requiring considerable professional input to turn them into courses, schemes of work and assessment strategies suitable for individual schools and pupils.

Geography is an optional subject in the 14–19 curriculum. For a brief period after 1991, it seemed that geography and the other foundation subjects would be compulsory at Key Stage 4, and the 1991 Orders contained programmes of study and statements of attainment relevant to this key stage. However, in 1994, ministers accepted Sir Ron Dearing's recommendations that there should be a more flexible situation at Key Stage 4, with a compulsory core of mathematics, English, science, modern foreign languages and technology, notionally filling 60 per cent of the time (mathematics, English and science only in Wales). Beyond this, schools would be left to make their own decisions about the availability of optional subjects and about the extent to which new vocational courses (GNVQ Part 1) should appear alongside academic options. As a result, the predominant way in which geography features at present is in GCSE courses as offered by the five GCSE examining groups in England and Wales, and the Northern Ireland Council for Curriculum and Assessment (CCEA). In 1995, geography ranked sixth in number of entries (see Appendix 3) still ahead of history, although newly designated core subjects like technology and languages may soon show an impact on this ranking for both subjects. At the time of writing, existing GCSE geography syllabuses are based on the original 1985 geography GCSE Criteria (DES/WO, 1985), but 1997 will be the last time these syllabuses are examined and a revised set of geography criteria have been prepared (February 1995). These will form the basis for all geography syllabuses taught from September 1996.

The revised GCSE Criteria (see 14–16 column in Figure 15.1) are consistent with the National Curriculum, requiring a similar balance of places, themes and skills and the need to address a range of scales of enquiry (DFE/WO, 1995). The specific mention of geographic issues and of attitudes and values provides a degree of progression from Key Stage 3 geography and also signals a welcome continuity from the original 1985 criteria. The latter were well received by the geography education profession and formed the basis for syllabuses which have, on the whole, had a very positive effect on teaching and learning in schools (see for instance Daugherty et al., 1991).

In the 16–19 curriculum, geography has always been an optional but popular subject. In 1995, it ranked seventh in subject entries with 43 454 candidates, although figures for the previous five years show that it is in continual competition with history (see Appendix 3). Geography is also available as an AS level, representing half the content but a similar standard to A level, although as with other subjects, these courses have not been popular. A and AS level geography syllabuses offered by the eight GCE boards in England and Wales and the CCEA in Northern Ireland, are, from September 1995, to fit the provisions of a Subject Core for Geography (SCAA, 1993a). The A/AS Subject Core, published in 1993, is an unusual document (see 16–19 column in Figure 15.1). Although described as a core, it does not provide a definable chunk of content which could form a recognisable part of all A/AS syllabuses. The requirements for study of 'a chosen physical environment', 'a chosen human environment', 'a theme which emphasises interaction between people and their environment' and 'personal investigative work' might better be described as an unusual kind of framework from which examination boards can develop syllabuses, leaving them considerable room for variation. Some of the emphases from earlier years are continued, for instance, the stress on physical–human relationships and skills of geographical enquiry. What is distinctive about the A/AS Subject Core is the focus on people–environment issues, on systems and on a greater breadth of specific skills and techniques. In many respects, the influence of the popular 'Geography 16–19' A level syllabus is more apparent in this Core than is any continuity with either the National Curriculum or the GCSE Criteria (both of which post-dated the Core).

This brief summary of the development and current state of geography 5–19 reveals the sequence of required content as set out in Figure 15.1, and provides an overview of the implicit progression. Certain aspects of continuity begin to emerge, including the need for courses to:

1. Maintain a balance between place study and thematic study (although this is less clearly specified in the A/AS Core).
2. Emphasise study of the interrelationships between physical and human geography.
3. Identify people–environment interactions and environmental concerns, and recognise that geography has a contribution to make to their understanding (even at Key Stage 1 from 5 to 7 years, with the inclusion of the environmental quality theme).
4. Develop pupils' abilities to undertake geographical enquiry.
5. Emphasise the application of geographical knowledge, understanding and skills to real world matters.
6. Ensure study at a range of scales, gradually expanding the emphasis from local to global.
7. Develop a framework of locational knowledge as the basis for understanding people, place and phenomena in geography. (In the National Curriculum, locational framework is specified on maps which accompany the Order as well as being referred to in the text.)

The wider context of curriculum change

While it is encouraging to note these features of continuity and agreement, it is important to sound a note of caution. In each case, the central guidelines are minimal only. Interpretation is crucial to ensuring continuity and quality in geographical education, whether it is in relation to teachers interpreting and expanding the new Geography Order, or to examination boards applying the Criteria and the Core to develop geography syllabuses. For many 5–14 teachers, it will not be easy to take on the required task of planning and developing their own curricula. The first version of the Geography National Curriculum (DES, 1991b; WO, 1991) seemed to require a reduction in professional input and the assumption of a 'delivery mode' rather than a 'creative mode' of operation (Lambert, 1994; Rawling, 1995). Although the broad geographical emphases of the 1991 Order were welcomed, as was the fact that it re-emphasised the contribution of geography to primary education, there were fundamental problems about its format and structure, many of which can be traced back to the lack of attention given by the Geography Working Group to principles of curriculum design (Rawling, 1992a). The Orders were overloaded and prescriptive, the 183 statements of attainment setting a record among National Curriculum subjects. The Working Group's interpretation of the attainment target and programme of study structure resulted in overlapping content, there was no distinctive key stage entitlement and no clear starting-point for curriculum planning. It was the recognition of these faults, highlighted in National Curriculum monitoring exercises (NCC, 1992; CCW, 1994), in reports from Her Majesty's Inspectorate (OFSTED, 1993; OHMCI (Wales), 1993) and in the work of the Geographical Association (GA) (Fry and Schofield, 1993; Rawling, 1992b) that led to the special case being made for geography in the Dearing Review. Paragraph 4.39 in the Dearing Final Report (SCAA, 1993b) allowed geography to restructure as well as to undertake the slimming which took place for all subjects. The 1995 Orders are the result and have essentially corrected the major structural faults.

However, for all teachers there will be a need for in-service support, exemplar materials and advice about new features such as level descriptions. Primary teachers, many of whom lack a specialist geography background, will need particular help. The summary of inspection findings for geography 1993/94 showed that there were significant weaknesses in primary geography teaching (OFSTED, 1995) and a small-scale monitoring exercise carried out by the SCAA (internal report) in the few months leading up to implementation of the National Curriculum, showed that both primary and secondary teachers perceived a need for more support, particularly in relation to assessment and level descriptions. The SCAA has already indicated that it will only be producing a limited number of guidance and information leaflets. A 'Geography and IT' leaflet was published in February 1995 (SCAA, 1995) and advice on using the levels at Key Stage 3 is expected in 1996. SCAA's position is that its statutory duty in presenting the new curriculum has now been completed and provision of further support will be left to the teaching profession itself. The GA with its newly emerging regional structure and active publication policy is well placed to provide support, but although materials and courses can achieve a great deal, it may not be so easy to address the loss of professional confidence and self-esteem felt by teachers. However, effective implementation of the 5–14 geography curriculum will depend primarily on the ability and willingness of teachers to reassert their professional roles as curriculum developers.

Of equal importance to the future well-being of 5–19 geography is the recognition that the emerging signs of continuity in geographical work only apply to the National Curriculum/GCSE/A and AS sequence of progression. In fact, beyond 14, geography is not only optional but also competing for attention in a much more dynamic curriculum context. At GCSE, although geography is a popular subject it will be competing for curriculum time against other National Curriculum foundation subjects like history, non-National Curriculum subjects like business studies, and possibly new vocational Part 1 GNVQ courses like leisure and tourism which are being piloted in schools in 1996/97. What is more, the possibility of subjects filling even smaller blocks of time in the 14–16 curriculum has been opened up by the agreement of ministers (June 1995) to allow short courses (5 per cent curriculum time) in a number of subjects including geography, and it remains to be seen whether this increased flexibility will help or hinder opportunities for geography.

Appropriate content for geography

What are the assumptions which underlie the sequence of content for 5–19 geography? How do higher education courses follow on from this?

For example, the emphasis on drawing together physical and human geography and ensuring that interactions are studied is explicit in the central guidelines right through from 5 to 19. Why is this the case? Livingstone (1992, p. 354) has referred to the way in which geography has frequently been seen as 'the integrating discipline par excellence that kept the study of nature and culture under one disciplinary umbrella' and he suggests that this go-between function has at various times 'had instrumental value in the search for institutional identity'. It might be argued that this lies behind the stress on physical–human interactions at school level. It is a survival strategy, aiming to maintain a coherence for school geography. It also serves to emphasise its links with sciences, humanities and the arts in order to strengthen its position in option columns, and to highlight its relevance to young people, who are interested in the kind of environmental concerns that cross physical–human boundaries. Geographers have been zealous in guarding both physical and human flanks of the subject in the recent National Curriculum discussions, as the controversy with science over who should teach earth science (Trend, 1994) and meteorology revealed. In higher education, it is usually the case that geographers see themselves as specialists in either a branch of physical geography or an aspect of human geography. Only rarely are voices in higher education heard to offer an overview, or to argue for bringing human and physical concerns back together in the interests of the unity of the subject (e.g. Johnston, 1991; Cooke, 1992). Should schools, therefore, be addressing broad environmental questions and issues, only selectively drawing on understanding of some aspects of physical and human geography and leaving basic physical and human geography to be addressed in higher education? Or should schools be providing a grounding in physical and human geography, with a gradual increase in the opportunities to apply this understanding to environmental matters, as pupils move from primary through secondary to higher education? How might this affect the content of first-year undergraduate courses and would there be implications for research activity?

O'Riordan (p. 114) is concerned that geographers may already be missing out on participating in key areas of environmental research, perhaps implying that it is time to reconsider the training and preparation of young geographers. Since the 1995 National Curriculum leaves decisions about breadth and depth of coverage to teachers, this is one area in which dialogue between teachers in schools and in higher education would be valuable.

Another set of assumptions about 5–19 geography seem to relate to a view of the way children learn. For instance, it seems to be implied that young children study places and topics at small scale and within their own experience first and then branch out to the unfamiliar and contrasting, increasing the scale of study as they do so. At Key Stage 1, study of the locality of the school is contrasted with a study of another locality, either within or beyond the UK. At Key Stage 2, it is clearly required that a locality from overseas must be studied and that pupils extend their ability to deal with regional and national scales. Key Stage 3 requirements extend to take in other countries, regions and localities outside the UK and to include international and global scales of study. Simple topics (rivers, factories, shops) are introduced at Key Stage 1 and more complex multifaceted topics and issues feature later (environmental change, settlement at Key Stage 2 and development, environmental issues at Key Stage 3). By the time A/AS level is reached, students should be able to investigate a complete people–environment theme emphasising interaction, different spatial scales, relevant processes, change through time and human responses. While this sequence appears to have a 'common sense' rationale to it, it is important to remember that there is, as yet, no clear agreement on identifying and measuring progression in geographical learning. Daugherty (p. 200) points out that one view of educational development (Egan, 1979) would challenge the generally held view of progressing from familiar to unfamiliar topics. In addition, it is surely not a simple sequence of expanding scales of study across successive age-groups. Even the National Curriculum Orders now recognise the need for pupils of 5–11 years to place their locality studies in a wider context (see Roberts and Healey—the zoom lens idea, p. 298), While many geographers including Jackson in this book (p. 91) have identified the need to understand hierarchies of scale as one of the fundamental challenges of modern geography.

Thirdly, some aspects of school geography are given considerable emphasis because they relate to current concerns or trends in society at large. Given general public concern about the environment, strengthened by the Rio Earth Summit in 1992 and the desire of the British government to show that it is taking action to address environmental issues, it is not surprising that geography should have been seen as an obvious National Curriculum subject to focus on the environment. This was certainly a factor in the decision of the original Geography Working Group to create an 'Environmental geography' attainment target, and this aspect of the subject was widely publicised by the media (Meikle, 1990). Some environmental education guidance planned for schools (DFE/DoE, 1995) is likely to give a strong steer towards the National Curriculum subjects of geography and science, despite the view from the educational profession that environmental education should permeate the whole curriculum and ethos of schools (British Government Panel on Sustainable Development, 1995).

Similarly, the idea that pupils should develop a basic framework of locational knowledge owes much to public concern voiced in the mid- and late 1980s. An International Gallup Survey report (1988) revealed that teenagers and adults in the UK were appallingly ignorant of their place in the world, and it was inevitable that attention should be focused on school geography to put this right. The original Geography Working Group devoted considerable time to a discussion of the locational knowledge needs of young people and, misguidedly as it now appears, wrote requirements for locational knowledge into the 1991 statements of attainment, as well as providing a series of locational knowledge maps. The Dearing Review ensured that both the maps and the locational knowledge requirements were more appropriately linked to programme of study requirements, but the need for such a framework remains and it would undoubtedly cause ministers some concern if it were to be removed. Both the GCSE Criteria and the A/AS Subject Core also state the need for an expanding framework of locational knowledge. Perhaps there should be a dialogue between schools and higher education about how such a locational framework might change over time. The National Curriculum is set for five years, but are the criteria of political and economic importance used to select places on the maps still likely to be relevant in the year 2000?

A further consideration relates to technological change. How might a locational framework be affected by the changing needs of an information-based society? The three trading zones of the world financial markets, brought into public focus by the collapse of Barings in March 1995, might be an equally or more appropriate division of the world to know about than the five continents, while access to the Information Superhighway will surely influence the kinds of skills and abilities appropriate to information gathering in an electronic age. Geographers have already been involved in consultation on the potential of the Information Super-highway (DFE et al., 1995). Such developments provide the stimulus for a major reconsideration of the kind of knowledge required by school-leavers and of the accessibility and availability of appropriate sources of information, whether these are paper maps, statistical databases, digital map data, on-line images or audio-visual links.

This brief discussion about the appropriate content at different ages should serve to draw attention to the need for schools and higher education to share criteria used in choosing content and to be more explicit about the thinking which lies behind them. Are there some general principles? How should the ways in which pupils and students learn affect the selection of content? How should the structure of the subject influence decisions about course characteristics? How far should the changing demands and interests of society be reflected in choice of content at any level—either for motivational or for more practical reasons?

Ensuring quality and continuity

The experience of implementing central curriculum guidelines for geography is beginning to raise many important matters for further dialogue. Two matters will be discussed here—the study of place and the development of geographical enquiry, both of which are fundamental to ensuring quality and continuity in geographical education.

The study of place

The development of the National Curriculum resulted in a strengthening of the position of place studies in school geography, after two decades in which geography had become dominated by thematic work and theoretical perspectives. The 1991 Geography Orders required the study of real places in their own right, and the Geography Working Group's clearly stated aim of developing pupils' 'sense of place' (paragraph 4.7, DES/WO, 1990) was well received by the profession. Unfortunately through lack of attention to curriculum structure, the original 1991 Orders (DES, 1991b; WO, 1991) succeeded in burying any notion of how to develop a sense of place under a weight of prescription of specific places and an overemphasis on facts and locational knowledge. In the revised Orders, the situation has improved. It is a requirement that places are studied at each key stage, but choice of examples is left to the teacher. Both the programme of study requirements and the level descriptions refer to the character of places, to distinctive features, to similarities and differences and to understanding the various scales of place study. This, coupled with the emphasis on active geographical enquiry and first-hand experience, provides the potential for re-emphasising places and for developing stimulating ways of understanding them.

However, the National Curriculum requirements are pieces of a jigsaw, and the responsibility for identifying the pieces and putting them together to produce a complete picture lies with the teacher. It is still worth remembering the two 'place study traps' identified by Johnston (1986) and to realise that there may still be a tendency to cover the required places as a unique catalogue of facts and information (the singularity trap) or to see places only as examples for thematic work (the generalisation trap). The latter is probably more of a threat given the new minimal requirements, particularly at Key Stage 3 where the long list of nine required themes seems to dominate over the briefer reference to two required countries. It may be a very real danger, for instance, that Japan is studied only as an example of economic success, that Brazil is merely the background for rain-forest studies, and that Bangladesh is seen as the example of a Third World country coping with flood hazard. The revised National Curriculum provides opportunities, but cannot ensure that teachers develop these in such a way that pupils appreciate the character and coherence of places in their own right, or that young people learn to recognise and express their own feelings about places.

At 14–16 and 16–19, the Criteria and Core are even less

explanatory about place study. The GCSE Criteria make it clear that a 'variety of places' at 'a range of scales' should be studied, but it remains for examination groups to translate this into syllabus requirements and to determine whether any syllabuses focus specifically on place studies or even require a real balance of place and thematic work. Of greater concern is the fact that the A/AS Core refers only to 'a human environment' with no clear explanation of what this means. Is it the same as a place? Elsewhere the requirements are for knowledge of places and for 'a balanced sampling of different spatial scales' 'embracing a range of geographical locations'. This specification suggests the choice of places to exemplify thematic work rather than a focus on places in their own right. Part of the difficulty in reviving the study of places in schools relates to teachers' lack of confidence and understanding of the opportunities and, possibly, to a fear of going backwards to old-fashioned 'capes and bays' geography. In higher education (as Johnston shows, p. 73), there is a considerable resurgence of interest in understanding places and a range of approaches from the exploration of perceptions, landscapes and cultural imagery to studies of the wider economic, social and political structures affecting places. The mid-1990s is the time to share ideas about the study of place. Not all of what is happening in higher education will be relevant to schools. Conversely, if none of the excitement is shared and if schools do not find ways of revitalising the study of place, then this is likely to have repercussions not only for continuity into higher education but for the way that future citizens understand the world they live in. The challenge for school geography is to ensure that the minimum central requirements are developed in ways that help pupils explore places as real entities, that engage the imagination as well as the memory, and that instil a sense of connections and links within and beyond the geography curriculum into their own lives.

The development of geographical enquiry

The need for pupils to participate and become competent in geographical enquiry is now specified throughout the 5–19 curriculum (see Figure 15.1) but the nature of the implied progression in geographical enquiry is not clarified. In the National Curriculum, each programme of study refers to enquiry questions (paragraph 1b) and to the range of enquiry experiences in which pupils should be involved (paragraph 2). There is some hint about progression in that simple enquiry questions and single-feature topics appear to be most appropriate at Key Stage 1 ('What is it like?', 'How did it get like this?'), while Key Stages 2 and 3 are progressively more concerned with 'How and why is it changing?', 'What are the implications?' and with complex topics like development and environmental issues. The GCSE Criteria and A/AS Subject Core both refer to the need for students to be able to use and apply a full range of enquiry skills and, for A level students in particular, to undertake personal investigative work.

The original Geography Working Group (DES/WO, 1990, p. 47) defined geographical enquiry as the active involvement of pupils in 'processes that enable them to find out more about the world and to explore and investigate geographical matters' and the current National Curriculum, Criteria and Core all refer to processes such as identifying questions, establishing sequences of investigation, collecting and analysing data, presenting conclusions and communication findings. However, it seems that we have little shared understanding of how, in practice, these processes of enquiry differ for 7, 11, 14 and 19-year-olds. What is different about the questions asked? What different teaching strategies are appropriate? What different outcomes are expected of pupil enquiry? Without agreement over these matters, we are in danger either of making the term so broad as to be meaningless, or of restricting it to a few teaching strategies like decision-making and resource-based problem-solving which are used repeatedly without a clear view of their purpose. Already it is apparent that there are misconceptions. Some examination boards seem to have taken on the term 'enquiry' to mean one major piece of coursework investigation rather than the underlying questioning approach to all work. Some textbooks appear to imply that enquiry work is always open-ended and pupil-directed, whereas a questioning, enquiring approach may apply on occasions to highly structured and teacher-led activities. Work undertaken with teachers by the SCAA (on Key Stage 3 exemplification of standards) has revealed how intimately interwoven the enquiry approach is with the knowledge and understanding requirements of the level descriptions and, consequently, how integrally linked are curriculum planning and assessment planning for National Curriculum Geography. The work is also beginning to highlight some key criteria for recognising progression in geographical enquiry, and this might usefully form the basis for discussions both across the 5–19 curriculum and with higher education.

The character and purpose of geographical education

The purposes of school geography, implicit in the official frameworks summarised in Figure 15.1, appear to be relatively narrowly interpreted and closely linked to the subject. However, many of the influential curriculum development projects in the 1970s and 1980s promoted the development of transferable skills and abilities like problem-solving, and communication skills, and addressed aspects of the personal, social and moral development of their pupils. The Geography 16–19 Project in particular (Naish et al., 1987) broke new ground in moving the emphasis, even at A level, away from purely subject-based objectives to a much wider view of geography as an educational medium for the 16–19 age-group, while the Geography, Schools and Industry Project (Corney, 1992) extended the experience of teachers and pupils in developing economic understanding and making links with business and industry.

All this appeared to change in the late 1980s when the National Curriculum was being established. The whole exercise was seen by many politicians as an opportunity to re-emphasise traditional subject knowledge and basic skills (Lawton, 1992). In 1991, the Secretary of State for Education, Kenneth Clarke, reinforced this view by responding to the NCC Consultation Report on Geography with strongly worded advice that 'the attainment targets should emphasise more strongly knowledge and understanding of aspects of geography, and put less emphasis on skills which, however desirable, are not particular to geography and less emphasis on the assessment of pupils' exploration of attitudes and values' (DES, 1991a).

When the Dearing Review of the National Curriculum was announced, many geographers saw it as an opportunity not only to correct the faulty structure of the 1991 Order but also to give a clearer signal about the purpose and character of geography at each key stage, and some hints as to its wider contribution to the whole curriculum. This was not to be, however. All the new subject Orders were required to follow a 'minimalist' line and focus only on specific subject requirements. The view of the SCAA was that there was no room in a Statutory Order for any reference to wider subject purpose, to teaching and learning approaches or to anything that sounded like guidance to teachers.

It therefore remains the case that teachers are left to make their own decisions about how (and, indeed, if) to use geography as a medium for broader educational objectives. The minimal nature of the revised Orders does not rule out such an approach and, indeed, it could be argued that by removing the weight of prescriptive content and giving teachers greater freedom of choice in curriculum planning and content selection, it makes such a task easier (Carter, 1994). However, given the past history of National Curriculum developments, there are many teachers who will need substantial support to move into a new mode of creative curriculum development, after several years of being de-moralised and deskilled by the original Orders. What is more, many supporting LEA and project networks have been dismantled, and schools are working in a more constrained environment where cross-curricular initiatives are less likely to be valued.

The implications of this situation should concern the whole geographic education community. The developments of the 1970s and 1980s showed that a broad view of the subject in schools, in which young people are actively engaged in exploring real issues relevant to their own lives, not only motivates and inspires them and so ensures the subject remains popular in schools, but also helps to develop the kind of skills, abilities and attitudes much prized by employers and higher education. The National Commission on Education (1993) also claims that transferable skills, flexibility, creativity and positive attitudes to change are key features essential for education in the next century. In other words, it is in geography's interests to take this broader approach, particularly given the growing competition from vocational courses which may seem, to pupils and parents, to have a more direct relevance to the world of work. Geographers in higher education can assist by making sure that teachers are kept up to date with the latest views about the relevance and application of modern geography. One area already identified is that of environmental education, where there is a need for a much clearer view of geography's potential compared with other subjects like science, and a shared understanding of the roles of school education and higher education (see Chapter 19). The DfE teacher guidance materials, already mentioned, might provide one stimulus for this debate. Another initiative, 'Pathways to the World of Work' is led by a group of industrialists and business representatives, who are co-operating to commission guidance on a range of cross-curricular themes (Blackburne, 1995). Recent developments in socio-economic geography and political geography might provide ideas for identifying local issues worthy of investigation by school pupils, and the increasing take-up of NVQs and GNVQs in schools and in higher education may stimulate greater debate about the work-related curriculum. The establishment of a minimum entitlement for the development of IT in schools (p. 224) also has clear implications for higher education, not all of which are necessarily being addressed, despite the existence of the Teaching and Learning Technology Project based at the University of Leicester.

Finally, discussion of the character and wider purposes of school geography inevitably raises the question of how this relates to geography in higher education. Does the situation suddenly change at 18 or 19, so that schools are seen as educating the whole person and higher education with imparting specialist subject knowledge and skills? Perhaps this never was a true description even when a university education was aimed at an academic élite; it certainly cannot reflect the situation in the 1990s when increasing numbers of students are enrolling for degree and degree level courses. Higher education must now address the educational needs of students with a wide range of abilities and motivations (see Chapter 16), and must cater for those who are using the subject as a step into more general employment, as well as those who are wishing to use their geography more specifically in research or teaching. Many of the articles published in the *Journal of Geography in Higher Education* are devoted to the broader educational objectives of geography (e.g. study skills, media skills). Healey (1994) has recently drawn attention to the way in which the assessment of teaching quality exercise in higher education has had the effect of refocusing attention on teaching and learning and on the wider purposes and outcomes of a degree course experience.

Certainly the criteria used in the Higher Education Funding Council (HEFC) exercise (original and revised 1995) make explicit reference to the aims of geography teaching and learning and to the needs of students, in language that has a definite educational, rather than subject, focus. Inevitably for many departments, this is leading to a reconsideration of the relative roles of teaching and research in higher education, and as Gardiner (1993) points out, the two may best be seen as linked and complementary rather than as competitive.

Overall, with schoolteachers wondering how best to expand their minimal geography entitlement 5–14, with new Criteria and Core about to be interpreted in GCSE and A/AS syllabuses, and with higher education teachers being directed to give greater attention to the quality and purposes of their teaching and learning, it would seem that the time is right for greater discussion and debate. How much can we share about the educational purposes of the subject? What scope is there for research into teaching and learning approaches and their effects on the broader educational development of pupils 5–19 and students 18+? How can we, together, ensure a high-quality educational experience as well as sound geographical learning? They are also the focus for the 1996 COBRIG seminar and it is hoped that this will give greater impetus to the debate and promote the development work that is undoubtedly needed.

Central control and professional freedom

The history of geography's place and character in the National Curriculum is essentially a story of changing curriculum control (Rawling, 1993). From a period of considerable freedom and teacher autonomy in the 1970s, we have moved through an era of substantial control and prescription epitomised by the 1991 Orders for Geography back to the current situation in the mid-1990s where, apparently, a greater amount of trust is invested back in teachers.

I have referred elsewhere to a number of criteria which seemed to characterise the period of innovation in geographical education in the 1970s and argued that these need to be re-created if school-based curriculum development is to flourish again (Rawling, 1996). The five conditions are:

What level?	Who does it?	What does it provide?
Level 1 General Level	National bodies Curriculum councils (e.g. SEAC, NCC, SCAA, ACAC) Curriculum projects (e.g. Geography 16–19, Avery Hill)	Broad aims Framework of themes, concepts, skills, etc. Procedures/guidance on interpreting the framework Assessment requirements and arrangements
Level 2 School Level	The school (primary) The geography department (secondary) [Examination boards undertake some aspects for GCSE and A/AS]	Specific objectives for geography in school and in each age-group Course outlines and teaching programmes Departmental strategy on assessment and on course review
Level 3 Classroom Level	The individual teacher in discussion with colleagues	Detailed schemes of work Detailed lesson plans and decisions about resources Plans for preparing and administering assessment tasks/tests Commitment to review and evaluate

Figure 15.2 Level of curriculum planning

1. A recognition of the different levels of curriculum planning and of the need to share control of the curriculum in an appropriate way (see Figure 15.2).
2. A recognition of the integral nature of curriculum and assessment.
3. Confirmation that it is possible to build on existing good practice in geographical education.
4. The provision of opportunities for teachers to work together in curriculum development activities.
5. A re-emphasis on the professional role and professional development needs of teachers.

The 1993 Dearing Review seems to have resulted in a more appropriate set of curriculum frameworks for the National Curriculum subjects, in which central direction has retreated to minimal requirements (level 1 in Figure 15.2), leaving schools and teachers the responsibility for details at the more specific levels of curriculum planning (levels 2 and 3 in Figure 15.2). The move towards level descriptions and the clarification of the way they link with the programmes of study, appear to give explicit recognition to the integral relationship between curriculum and assessment. In addition, the revised Geography Orders confirm a number of features of past good practice, such as key ideas, geographical enquiry and the relevance of geography to environmental, social and political issues. However, the extent to which teachers find time, enthusiasm and opportunity to work together, and the degree to which their professional development needs will be met in the new, more market-oriented, situation of the late 1990s remain to be seen. The central curriculum and assessment authorities (SCAA and ACAC) do not have the remit to provide much further advice and support. Local education authorities have changed in character and are no longer able to offer the kind of curriculum support common in previous decades, and restrictions on education spending are causing concern throughout England and Wales in 1995. Schools are often too busy making their own arrangements for managing budgets and resources to give much attention to subject-specific curriculum development issues.

For a subject like geography, the professional support, essential to ensuring that good quality geographical education arises out of the minimal framework, is most likely to come from bodies such as the Geographical Association (GA) and the Royal Geographical Society with the Institute of British Geographers (RGS/IBG). Effective support should include not only curriculum planning advice but also professional dialogue about geography throughout the schools/higher education continuum. The GA's new regional network provides an obvious channel for promoting teacher teamwork and in-service activity, in co-operation with what remains of LEA advisory services and the GA's own branches. The greater dialogue with higher education being encouraged by the Council of British Geography and by the new RGS/IBG Education Committee could ensure a much needed input of subject expertise to teachers of the 5–14 age range. A period of renewed innovation in school geography will only occur if the National Curriculum and developments in 14–19 geography are seen as an opportunity to reopen exchange with academic geographers and to start talking about the subject again in a constructive way.

The format of central guidelines raises another set of issues. For 5–19 geography, there is a National Curriculum, a set of GCSE Criteria and an A/AS Subject Core. Why is one kind of framework appropriate at one age-group and not at another? Or is this just a reflection of the time at which the frameworks were developed? If there are to be central guidelines, it might be argued that they should follow, at least broadly, a similar format. More significantly, what should central guidelines contain?

The National Geography Curriculum is effectively a list of content requirements. The official line has been that a Statutory Order should only give the minimum content (the 'what to teach') and that the more specific educational objectives and teaching/learning strategies (the 'how to teach') should be left to schools (Tate, 1994). The reluctance of SCAA or the DfE to promote cross-curricular themes and dimensions may be seen to reflect this view. However, it could be argued that many current trends in society seem to be leading to a rather different view of the most appropriate characteristics for an educational framework for the twenty-first century. The National Commission on Education (1993) quotes Drucker and draws attention to the importance of knowing how to use and apply knowledge. Drucker refers to 'knowledge workers'—people who know how to put knowledge to good use (Drucker, 1993). The National Commission suggests that there are two key characteristics of high-quality education—one is that high-quality and constantly updated knowledge and skills should be incorporated into the *content* of the curriculum and the second is that a certain *style of teaching and learning* is required.

> Good teaching will foster in students a spirit of inquiry about the world around them. It will encourage them to think for themselves, to be critical and to be self critical. It is people who make the world what it is and every young person has the opportunity to change it.
>
> Education is about empowerment as well as the transmission of knowledge (Report of National Commission on Education, 1993, pp. 39–40).

The first version of the National Curriculum overemphasised and over-represented the content dimension of the curriculum, and under-represented the importance of teaching and learning approaches. It might be argued that Dearing has put the first problem right by reducing overload, but has still not addressed the second, and that an educational framework fit for the twenty-first century should contain specific reference to the need to foster critical enquiry, to the importance of attitudes and values and to the appropriate range of teaching and learning opportunities. It is interesting to note that one of the issues thrown up by the Dearing Review of 16–19 qualifications is the extent to which a broader definition of core skills for all 16–19-year-olds should include aspects like problem-solving and critical enquiry (Dearing, 1995). Despite the fact that his brief focused on qualifications, the final Dearing 16–19 Report may well make recommendations which will have significant implications for future styles of teaching and learning.

Finally, of course, the discussion of central guidelines at school level raises the question of whether there will be moves to suggest a core curriculum for geography degree courses (see Bradford, p. 285). Although academics have distinctive concerns that are not shared by schoolteachers—the close link between teaching and research, for instance—there may be value in exchanging experience in relation to central control and professional freedom.

Conclusion

The recent period of change in school geography has led to a situation in which there are now central guidelines, in some form, covering every age-group from 5 to 19-year-olds. There is a degree of curriculum continuity planned into the system, particularly at 5–14, but a close scrutiny of the sequence of content 5–19 throws up more questions than answers about progression in teaching and learning geography. We are still a long way from understanding how pupils develop understanding of key concepts in the subject, how they are best helped to integrate physical and human aspects of geography, how they develop a 'sense of place' or what constitutes progression in geographical enquiry. What is more, the central guidelines, whether National Curriculum, Criteria or Core, provide only the outline framework of an entitlement to geography, and much is dependent on teachers, examination boards, publishers and in-service providers to expand these guidelines into sound experience for pupils. Given the lack of trust displayed over the past 10 years or so, there is also need for rebuilding teacher confidence and regenerating enthusiasm in order to undertake this task.

All these issues are of crucial importance to geographers in higher education. Unless schools do develop courses which stimulate and motivate young people, the supply of geographers into higher education will dwindle and decline in quality. The task will be made easier if teachers can share some of the deeper understandings and potential excitement derived from developments at the research frontier. But beyond this joint concern with survival, lie the deeper shared questions of what a geographical education can contribute to the preparation of future citizens, and how this is best promoted for different age-groups. If geography is to move into a new era of innovation and strength in the changing educational context of the next century, then it is vitally important that these questions have been debated, and that some mutually acceptable answers have been found.

References

British Government Panel on Sustainable Development (1995) *First Report of the British Government Panel on Sustainable Development*. Department of Environment, London.

Blackburne, L. (1995) Business rescues cross-curricular themes. *News, Times Educational Supplement*, 6 January.

Corney, G. (1992) *Teaching Economic Understanding through Geography*. Geographical Association, Sheffield.

Carter, R. (1994) Feet back on firmer ground. In Geography Extra, *Times Educational supplement*, 18 November, Times Newspapers Ltd, London.

Cooke, R.J. (1992) Common ground, shared inheritance; research imperatives for environmental geography. *Transactions of the Institute of British Geographers*, New Series, **17**(2), 131–151.

CCW (Curriculum Council for Wales) (1994) *Monitoring Report on National Curriculum Geography in Wales, 1993*. CCW, Cardiff.

Daugherty, R., Thomas, B., Elwyn Jones, G. et al. (1991) *GCSE in Wales; a Study of the Impact of the GCSE on the Teaching of History, Geography, and Welsh*. HMSO, Cardiff.

Dearing, R. (1995) *Review of the 16–19 Qualifications Framework: Summary of the Interim Report. The Issues for Consideration*. Central Office of Information, London.

DES (Department of Education and Science) (1991a) Letter accompanying *National Curriculum: Geography Draft Statutory Order*, DES, January 1991.

DES (1991b) *Geography in the National Curriculum (England)*. HMSO, London.

DES/WO (Department of Education and Science and Welsh Office) (1985) *GCSE, The National Criteria: Geography*. HMSO, London.

DES/WO (1990) *Geography for Ages 5–16: Proposals of the Secretaries of State for Education and Science and for Wales*. HMSO, London.

DfE (Department for Education) (1995) *Geography in the National Curriculum (England)*. HMSO, London.

DfE/DoE (Department for Education and Energy Efficiency Office of Department of Environment) (1995) *Education and the Environment, the Way Forward: Conference Report*. HMSO, London.

DfE/WO (Department for Education and Welsh Office) (1995) *GCSE, The National Criteria: Geography*. HMSO, London.

DfE, WO, Department of Education Northern Ireland, The Scottish Office (1995) *Superhighways for Education; Consultation Paper on Broadband Communications*. HMSO, London.

Drucker, P.E. (1993) *Post-Capitalist Society*. Butterworth Heinemann, London, p. 170.

Egan, K. (1979) *Educational Development*. Oxford University Press, New York.

Fry, P. and Schofield, A. (1993) *Teachers' Experiences of National Curriculum Geography in Year 7*. Geographical Association, Sheffield.

Gallup Organisation (1988) *Geography: An International Gallup Survey* (Report for the National Geographic Society, USA). Gallup, Princeton, NJ.

Gardiner, V. (1993) Teaching, learning and research—on the separation of the indivisible. *Journal of Geography in Higher Education*, **17**, 180–186.

Healey, M. (1994) Geography and quality assessment. *IBG Newsletter*, 2–4.

Johnston, R. (1986) To the ends of the Earth. R.J. Johnston (ed.) *The Future of Geography*. Methuen, London, pp. 326–338.

Johnston, R. (1991) *A Question of Place*. Blackwell, Oxford.

Lambert, D. (1994) The National Curriculum: what shall we do with it? *Geography*, no. 342, 79(1), 65–76.

Lawton, D. (1992) *Education and Politics in the 1990s; Conflict or Consensus*. Falmer Press, London.

Livingstone, D.N. (1992) *The Geographical Tradition*. Blackwell, Oxford.

Meikle, J. (1990) Teach pupils to save world. In Today Column, *Daily Mirror*, 7 June, Mirror Group Newspapers, London.

Naish, M., Rawling, E.M., Hart, C. (1987) *Geography 16–19: the Contribution of a Curriculum Project to 16–19 Education*. Longman for SCDC, Harlow.

National Commission on Education (1993) *Learning to Succeed, Report of the Paul Hamlyn Foundation National Commission on Education*. Heinemann Educational, London.

NCC (National Curriculum Council) (1992) Implementing National Curriculum Geography. Unpublished report of the responses to NCC Questionnaire Survey, NCC, York.

OFSTED (Office for Standards in Education) (1993) *Geography Key*

Stages 1, 2 and 3. The First Year 1991–1992. HMSO, London.

OFSTED (1995) *Geography, A Review of Inspection Findings 1993–1994.* HMSO, London.

OHMCI (Office of Her Majesty's Chief Inspector) (1993) *A Survey of Geography in Key Stages 1, 2 and 3 in Wales.* HMSO, Cardiff.

Rawling, E.M. (1992a) The making of a National Geography Curriculum. *Geography*, no. 337, 77(4), 292–309. Geographical Association, Sheffield.

Rawling, E.M. (1992b) *Programmes of Study: Try this Approach.* Geographical Association, Sheffield.

Rawling, E.M. (1993) School geography: towards 2000. *Geography*, no. 339, 78(2), 110–116.

Rawling, E.M. (1996) The impact of the National Curriculum on school-based curriculum development in secondary geography. In *Geography in Education.* Cambridge University Press, Cambridge.

SCAA (School Curriculum and Assessment Authority) (1993a) *GCE A and AS Subject Core for Geography.* SCAA, London.

SCAA (1993b) *The National Curriculum and its Assessment; the Final Report of Sir Ron Dearing.* SCAA, London.

SCAA (1995) Geography and information technology. In *Key Stage 3: Information Technology and the National Curriculum.* SCAA, London.

Tate, N. (1994) Target vision. In Platform, *Times Educational Supplement,* 2 December, Times Newspapers Ltd, London.

Trend, R. (1994) Rocks and roles. In Geography Extra, *Times Educational Supplement,* 19 November, Times Newspapers Ltd, London.

Welsh Office (1991) *Geography in the National Curriculum* (Wales). HMSO, Cardiff.

Welsh Office (1995) *Geography in the National Curriculum* (Wales). HMSO, Cardiff.

Part D

Learning from the dialogue

Geography at the secondary/higher education interface: change through diversity

16

Michael Bradford

This chapter addresses changes in general in secondary and higher education before discussing changes specific to geography. These more general changes provide a context for the changing interface between the secondary and higher education sectors for all students. It may help to conceptualise the interface. It can be seen in terms of communication between the staff of the two sectors, as the point at which students move from one sector to another, and potentially as an area of smooth progression in learning methods and subject content. Currently, however, the interface more often presents a gap or discontinuity in methods and content, and it is this which the chapter seeks to explain. Academic geographers and secondary teachers, who seem to communicate less than they did in the past, need to recognise the gap, partly so that students can be helped to bridge it and partly to ensure the future health of the subject. A major question which needs to be addressed by both sets of people is the degree to which there should be uniformity or diversity in what is learned within each of the two sectors, which in itself affects the character of the interface.

Geography into the Twenty-first Century, Edited by E.M. Rawling and R.A. Daugherty,
© 1996 John Wiley & Sons Ltd.

Changes in secondary and higher education

There has been considerable change in both the secondary and higher education sectors since the mid-1970s. In the secondary sector there is now a range of institutional provision for 16–18-year-olds, including 11–18 schools, and sixth-form and tertiary colleges. The private/state divide in educational provision, characteristic of the late 1970s, has become in some areas a continuum with grant-maintained schools and city technology colleges occupying the middle of the continuum and private schools with assisted places complicating the private end of the continuum (Bradford, 1993, 1995a). So the types of establishment and context in which 16–18-year-olds study geography have diversified.

There has always been considerable variation within both the state and private sectors, but these recent changes have further increased the differences in, among other things, class size, resources, and range of other subjects available for study alongside geography. In small schools, students may be taught by the only member of the geography department, who is then responsible for making any selection of options within the course. In large colleges, there may be a range of specialist staff available who can offer students a choice of options within their course. Staff in 11–18 schools have been coping with the National Curriculum and more recently with its slimmed-down version, as well as with the General Certificate of Secondary Education (GCSE) and sixth-form studies. Staff in tertiary or sixth-form colleges have been able to devote less divided attention to changes in A level and AS, and to a series of more vocationally oriented qualifications at this level, the latest of which are General National Vocational Qualifications (GNVQs). Some students, therefore, experience a considerable degree of continuity in their staff and fellow students between the ages of 16 and 17; while others face new environments, new staff and fellow students who have come from different educational contexts and have varied backgrounds in geography. For some the school-to-college move helps the transition to higher education.

The context in which teachers work has also changed, in many cases dramatically. Teachers in many grant-maintained schools, for example, were isolated from local education authority (LEA) schools and advisory services (Bradford, 1995a) when the first version of the National Curriculum was being implemented in the late 1980s. Since that time, in many LEAs, the advisory service has been depleted, almost disappeared or what remains plays an inspecting role which does not encourage the seeking of advice. For these reasons, too many teachers were lacking support and were, in many ways, 'reinventing the wheel' in course planning and resource preparation. This very variable help for teachers presents the Geographical Association (GA) and similar bodies with an opportunity and a challenge to fill the gap, but one can only deplore that there is a gap there to fill. Teachers in most schools also face cuts in funding for new resources which have made the introduction of the National Curriculum and new syllabuses even more difficult. No wonder that they have had little opportunity to keep abreast of changes in higher education.

In the higher education sector, a diverse situation has become less so in some ways. Polytechnics have become universities. They are funded by the same bodies as older universities, the Higher Education Funding Councils for England (HEFC(E)) and for Wales (HEFC(W)). Their teaching and research are rated in the same way. Differences still exist in resource levels and student–staff ratios, but the older universities now share the newer universities' concern with teaching large classes. Many have moved to semesters and modular programmes, although it should be noted that the word 'modular' covers a multitude of sins or attributes, depending on your point of view. There seems to be a move, however, to produce a hierarchy of universities, with varying emphases on research. For some universities, research and postgraduate teaching may have much greater prominence; while there is a threat that some others will become teaching-only institutions. The superficial unifying tendency of recent years may simply be a prelude to further diversification.

At present, however, students entering higher education still have institutions with different sizes and locations from which to choose, a range of size of geography departments, and different types of programmes with geography as the single, joint or part element. Some have more choice than others. The financial support of students is much diminished, so that some feel they are restricted to universities accessible from the parental home. Many run up large overdrafts by the end of three years. Finance and diet have become two areas which today's students have to learn to organise along with their time, in order to reduce the stress of university life. Some have more guidance over university entrance than others. Some can afford to visit many institutions before making their selection. Others have to choose, sight and site unseen. Wherever they finally go, they will usually find large first-year lecture classes, much larger than those which they have experienced at secondary level. For some, even their university small group teaching will be in groups larger than the secondary level classes which they have left. In some institutions they will meet in small classes less than once a week or even less than once a fortnight. Some departments will balance intensive (small group) and extensive (large lectures) teaching throughout the three years. Others will leave the intensive work until the final year, by which time it is sometimes too late. A few will front-load the intensive elements to act as a bridge from secondary to higher. The gap has to be recognised by both staff and students, and attempts made to bridge it. When arriving in higher education, students can no longer rely on study skills acquired at the secondary level. They have to learn some new ways of studying and they can no longer rely on the level of access to staff that their teachers enjoyed. The pressures of research and of larger numbers mean that one of the major challenges facing university departments is establishing ways of devoting the necessary individual attention to all students, not just to those who seek it, to ensure that they have sufficient opportunity to realise their potential.

Relations between secondary and higher education in geography

With these more general changes as a background, it is now possible to view the relations between secondary and higher education in geography. Since the mid-1970s the two sectors have grown apart. This is partly because of the above pressures for change within each sector. There is less contact within bodies like the GA, because higher education geographers are less involved than in the past. Research has been the prime criterion for promotion, and service to the discipline in other ways has counted for little. Perhaps this is changing with the attention given to learning in the recent round of teaching quality assessment and with the need for universities to review their promotional criteria where staff profiles with varying degrees of research involvement are encouraged. At present though, there are relatively few academics actively involved in secondary education. This means that there is considerable ignorance about the changes that have been going on there. Few would know the structure of the National Curriculum, unless they have children at school. Still fewer would know the structure of the A/AS Subject Core for geography though, to be fair to higher education colleagues, relatively few secondary teachers knew of this until early 1995, which says much about the degree of communication of imposed change. Academics may have criticised some of their students for the poorer quality of their essay-writing skills, without knowing that in some A level syllabuses the emphasis is on resource-based questions and that less essay writing is required. On the other hand, most know that there is more coursework carried out within schools than in the mid-1970s and that students are used to having it included within their overall assessment.

In higher education, the emphasis until recently, especially in the older universities, has been on the content of courses and perhaps the overall structure of the programme. Few thought in curriculum terms. Yet geography is one of the few subjects with its own *Journal of Geography in Higher Education* which has a strong, but no longer an exclusive, orientation of improving learning and teaching. The small original core of geographers, mainly from the former polytechnics, with an interest in innovation in teaching and learning has grown and widened, but still there is less interest in pedagogy than in secondary education. So among those teaching geography in higher education, there is ignorance of secondary education, and within their teaching, a different balance of interests than in secondary education.

The content of degree programmes in geography varies greatly. Many reflect the research interests of the staff, especially in the final year. Some, more than others, reflect innovations, such as new cultural geography or geographical information systems (GIS), because of the availability of staff and/or resources. An overview of university programmes would probably conclude that there was growing diversity, because during the 1980s and 1990s there has not been one major trend affecting as many areas of the subject as did either the scientific revolution in the 1950s and early 1960s, or the radical geography movement in the early 1970s. The absence of such major changes may partly account for the reduced impact of higher education on the geography taught in secondary education.

A view of secondary education since the mid-1970s suggests that it has been more concerned with pedagogy than with the content of the subject, at least until the National Curriculum, which was initially highly content-driven. The subject at the secondary level seemed still to be strongly affected by the scientific revolution of the late 1950s and early 1960s. In GCSE and A level courses, hypothesis testing abounded and still does, in many cases as the sole form of investigative work. Locational analysis and models prevailed in many syllabuses (and remarkably still do in some), which perhaps as a result are highly economic in their view of human geography (Bradford, 1995c). Yet physical geography, where positivist approaches are more appropriate, was, until the National Curriculum, poorly represented in many GCSEs and in some A level syllabuses. This was a focus of complaint for many in higher education, some of whom decided to mount remedial classes for those who had taken the highly innovative Geography 16–19 syllabus. This incorporated an enquiry approach to learning and was probably more innovative in terms of pedagogy than content, though it did lead the way in placing much emphasis on a people–environment approach. Until very recently this was the 'new syllabus' and yet it was devised in the late 1970s.

Perhaps it should not be surprising then that new teachers direct from university geography courses have found that many of the A level syllabuses that they are required to teach are far removed from their experience as far as parts of the recently introduced Northern Examinations and Assessment Board (NEAB) A and AS syllabuses seemed initially to teachers in their forties and fifties. It can be argued that the inclusion in these NEAB syllabuses of some aspects of social and political geography, to balance the economic and technological processes, gives those leaving school and geography a broader geographical preparation for life. For those continuing in geography it also provides a more complete background for university. The inclusion of crime and disease among hazards seemed a major change to some, but both are inherently spatial and environmental, and disease, at least, has been taught elsewhere within schools for decades. With so much change elsewhere in the curriculum, some teachers may have become, understandably, too set in their A level ways. Unfortunately holding or improving positions in league tables may be hindered by the uncertainty of changing syllabuses. Yet some of even the present A and AS syllabuses have too little relevance to the changing world, and are by no means as good a preparation for life or future study as they could be. So, much needed to be done to improve A level syllabuses before the recent introduction of the A/AS Subject Core.

One of the positive attributes of the National Curriculum, as seen from higher education, was to reinforce geography's position as a bridging subject between earth sciences and social sciences. The balance in the National Curriculum between physical geography, people and their environments, and human geography has also been re-emphasised in the A/AS Subject Core for geography, which all syllabuses now have to include. The first group of syllabuses recognised as satisfying the Subject Core requirements, began in 1995 with the first terminal exams for A level in 1997. The syllabuses also have to conform to the new Code of Practice which incorporates the School Curriculum and Assessment Authority's (SCAA) modular rules and the 1992 Principles of one of its predecessors, the School Examinations and Assessment Council (SEAC). The Principles, among other things, restrict coursework to a maximum of 20 per cent. The Subject Core is flexible and is not likely to stultify the subject in the way that the original National Curriculum might have done. Yet its emphasis on one physical and one human environment is not well received in higher education. Principles of physical geography, for example, need to be established, and this is difficult to accomplish through the study of one environment. Although it may be argued that comparisons may be included in the non-Subject Core part of the syllabus, as it stands the Subject Core does not establish as good a 'model' for planning geography courses as it might. Most of the syllabuses, now approved, were developed at great speed (Bradford, 1995c). There is still much to be done to present an appropriate range of syllabuses which are not set in their time, in many cases, well before the 1990s. Therefore, one challenge at the secondary side of the interface is to improve provision at A level.

The emphasis on one environment in the Subject Core seems to reflect a desire, which is also seen in the National Curriculum, to return to real places and to swing the emphasis back to place and away from space. Although there has been considerable work on the 'sense of place', this is not the aspect of place that is conveyed in the National Curriculum. Nor is it the interdependence of places and the global–local relations that concern so much of higher education research and teaching. In the original National Curriculum the specification of place study seemed too much like traditional regional geography, with understanding sought solely within the place. The slimmed version of the National Curriculum puts greater emphasis on the interdependence of places and on seeing them in a broader context. However, nothing could be added in the slimming of National Curriculum Geography, the content was simply reorganised, and so there is still little to stimulate studies of the 'sense of place'. Much still needs to be done to produce a balance between space and place, and indeed a broadening of the types of investigations that go with it. Qualitative methods, for example, though mentioned in the Subject Core, do not seem to be present in many of the new syllabuses. Quantitative methods still dominate most syllabuses.

Diversity or uniformity?

The A/AS Subject Core in geography has not produced a continuation of the National Curriculum in the sense that everyone will arrive at university or college with the same background. There will still be a wide range of backgrounds, made even wider by the range of new GCSEs that will begin in 1996 with first examination in 1998. These will fill the void left by geography being excluded as a statutory subject from Key Stage 4. Backgrounds in physical geography should be better in overall terms but, given the flexible nature of the mostly modular A level syllabuses, some students may have completed their requirements in physical geography long before arriving in higher education. Some academics may decry the continued variation in backgrounds, but still more would be opposed to a uniform background that neither fulfilled their needs nor changed over time. The flexibility of the Subject Core and the range of syllabuses permit varying approaches to geography and to learning, rather than presenting a stultifying uniformity which invariably would be dominated by one approach to both the subject and learning. Such an approach would prove no doubt to be very much set in its time or, if the original National Curriculum is anything to go by, set in a previous time.

Advocates of a unified background forget that many students entering higher education are 'mature students', who come from Access courses or come with other qualifications such as those of the Business and Technical Education Council (BTEC) or, in the future, with GNVQs. Not only is the subject content of courses[1] related to these qualifications different, the approach to learning and the forms of assessment are also very different. A level reflects the long term influence of the Department for Education (DfE), while more vocationally oriented qualifications reflect the influence of the Department of Employment (seen also in universities through Enterprise in Higher Education). Their differences are epitomised in their contrasting views of coursework (Bradford, 1995a) and it is too early to predict the likely consequences for education of the merger, announced in July 1995, of those two departments.

The advent of the A/AS Subject Core, or 'Key Stage 5' as some refer to it, along with a misplaced view that it will make backgrounds more uniform, may increase the impetus for a common core in the first year of higher education, or 'Key Stage 6'. Reduced human and material resources, the profit motive of publishers, and the technological drive of the Teaching and Learning Technology Project (TLTP) may all be considered pressures in the same direction (Bradford, 1995b). Even the compulsory training element for post-graduates, imposed by the Economic and Social Research Council, and which is a model for another section of higher education (Bradford et al., 1994), could also emphasise the trend towards uniformity of practice.

Yet the backgrounds of students coming from secondary education will still be very varied. The first year in higher education does not just build upon what has gone before. It is as much a base for what follows and will be influenced by this. Given the general move to research clusters within departments, in terms of concentrations of sets of specialists, rather than a broad coverage of the subject, what follows the first year has become more diverse. It is less likely than ever that students will emerge from higher education with similar backgrounds. This will continue to present a major problem for teacher training. The variable delivery of geography, as single honours, joint honours, combined studies or as part of a cafeteria-style modular degree again make the implementation of a common core for geography in higher education more difficult. Some of the dangers inherent in establishing a common core for geography in higher education are similar to those which have affected the National Curriculum and the A/AS Subject Core. It might be open to the influence of a few, be set in its time and be difficult to change. It might also prove a very easy target for political interference (Bradford, 1995b). Perhaps it is better to retain the flexibility to incorporate change, to present students with a real choice of courses and to maintain differences between universities. There are plenty of opportunities for people to share resources through TLTP and the World-Wide Web, if they fit their programmes. Whatever the future for the first year, it is essential that geography is presented to students in coherent course units, whether as a single honours programme or as part of a modular, multi-disciplinary degree.

In summary, it seems that, in the near future, students entering higher education from secondary education will come from varying educational institutions associated with very different backgrounds of learning. They will also continue to arrive with varying backgrounds in geography and to receive a diversity of programmes in geography.

Bridging the gap

A major future concern for geographers will be the bridging of the gap that has emerged between secondary and higher education. Although there are wider issues which are difficult for geographers alone to influence, there is much that geographers can do. The sectors can learn from each other. Higher education can benefit from learning about curriculum development in secondary education (at least as developed pre-National Curriculum), about approaches to learning and about student profiling and assessment. Secondary education could profitably be updated about research trends in the subject and, indeed, about what is taught in higher education. Some of the ideas may be suitable for inclusion in secondary education. What has to be avoided are two extreme views. There are a few in higher education who think that all secondary education should do is prepare students for them—'let them provide the background and leave the interesting stuff to us'. There are also some in secondary education who think that those in higher education just wish to establish a hierarchy down which their ideas filter, so that they can control what is taught in schools. But there are much more positive scenarios than this with, for example, two-way communication on the appropriate levels and ways in which concepts and ideas can be learned.

There is a great need for conferences which provide opportunities for updating and interchanges between teachers in the two sectors. A national seminar, such as the one the Council for British Geography (COBRIG) held at Oxford in 1994, can only include a relatively small number of people; more locally based conferences are also needed. Some university departments mount meetings as part of their promotional activities. While hearing about recent research is useful to teachers, particularly when oriented to syllabuses, it would be useful if such meetings were organised more as two-way events with some teachers also leading sessions on developments in schools, such as GNVQs, new syllabuses or new ways of monitoring students' progress. These may be useful to secondary colleagues as well as to academics seeking an understanding of the background of their future students.

Newly and recently trained teachers can form a very useful resource and agent of change. They may be well qualified to prepare packages of new material, addressing emerging areas of the subject, for publication through such organisations as the GA. Properly accredited courses for teachers with a few years experience could be set up for them to share approaches and develop materials. Such courses could further their enthusiasm and help to sustain a strong body of committed geographers.

Conclusion

General sectoral change has been rapid with too little time to adapt and consolidate. Pressures within each sector have undoubtedly contributed to the gap widening between them. This gap affects individual students who every year try to bridge it. Both sectors have a responsibility to their students to try to reduce the gap and attempt to find ways of helping them to bridge it. For the future health of the subject, and to ensure its continued popularity in secondary level and the continued flow of geography students from secondary to higher education, members of both sectors have an obligation to reduce their mutual ignorance. They need to learn from each other, to improve learning and teaching environments in both sectors, and to produce a varied set of balanced, coherent curricula for both sectors. This author prefers those changes to be through diversity rather than uniformity.

Acknowledgements

I would like to thank the other members of a seminar group at the COBRIG conference in Oxford, July 1994 for their discussion, which was summarised by Andrew Powell of Kingston University.

[1]Access courses are pre-entry programmes for mature entrants to higher education.

References

Bradford, M.G. (1993) Population change and education: school rolls and rationalisation before and after the 1988 Education Reform Act. In T. Champion (ed.) *Population Matters: the Local Dimension*. Paul Chapman Publishing, London, pp. 64–82.

Bradford, M.G. et al. (1994) Searching for good practice in postgraduate education: Arena symposium. *Journal of Geography in Higher Education*, **18**, 335–372.

Bradford, M.G. (1995a) Diversification and division in the English education system: towards a post-Fordist model? *Environment and Planning A*, **27**, 1595–1612.

Bradford, M.G. (1995b) A common core or food for the political worm: editorial. *Journal of Geography in Higher Education*, **19**, 5–9.

Bradford, M.G. (1995c) A review of the new A level and AS syllabuses. *Teaching Geography*, **20** (3), 145–154.

Human and regional geography in schools and higher education

17

Mick Healey and Margaret Roberts

Introduction

The people who are probably most acutely aware of the gap between human and regional geography in schools and human and regional geography in higher education are recent geography graduates who have specialised in these areas and are following Postgraduate Certificate in Education (PGCE) courses to become secondary schoolteachers. They are at the interface. Their understanding of the subject as pupils has usually been transformed during their undergraduate courses. During their PGCE courses they have to return to what counts as human and regional geography in schools. It is important that they ask critical questions about that interface, because it is they who will provide the next generation of teachers whose own views of geography will influence the character and quality of the subject in schools.

An informal survey of one such group of geography graduates on a PGCE course, training to become secondary school geography teachers, revealed a variety of expertise in human and regional geography at undergraduate level. Sam did a detailed course on the geography of elections in the UK. John enjoyed studying Chinese landscape paintings as a way of understanding different cultural perceptions of the landscape. Alison studied the meaning of urban landscape through the study of graffiti in Paris. Danielle studied poetry

Geography into the Twenty-first Century, Edited by E.M. Rawling and R.A. Daugherty,
© 1996 John Wiley & Sons Ltd.

and literature as a way of understanding cities of Europe. She also studied 'sex tourism' in South East Asia. Robin studied patterns of joy-riding. Andy studied perceptions of the East End of London gained from *EastEnders*. Richard studied the allocation of council houses and the right to buy. Rachel studied land resettlement in Malaysia through in-depth interviews. What happens to this kind of expertise in human and regional geography when these student teachers start teaching?

They are not likely to teach any of those themes, nor use those approaches in their geography teaching in secondary schools. Of course, during their undergraduate courses, most have studied themes and places which are directly relevant to the school curriculum, but they are aware of the gap between what they have just experienced and what they are required to teach. They have encountered not only different topics of study, but also different approaches to geographical knowledge, different ways of studying places and different ways of being taught. It is worth examining the differences more closely before discussing the issues arising from them.

Differences between schools and higher education

The structure and content of human geography

Human geography in schools

Three characteristics of the present state of human geography in schools can be identified: the dominance of human geography over other aspects of geography; the way it is structured through themes; and the emphasis on economic and social aspects of the subject. All these characteristics are evident in a summary of content to be studied for General Certificate of Secondary Education (GCSE) examinations for one of the four examination groups in England (Table 17.1). Although there is some variation between examination groups, these five syllabuses are reasonably representative of what can be studied by 14–16-year-olds for GCSE geography examinations to be taken in 1996.

Human geography has had a prime place in school

Table 17.1 Themes listed in NEAB's syllabuses for GCSE examinations in 1996 (themes almost entirely in the human geography domain in bold, themes with an asterisk include some human geography topics)

NEAB A	NEAB B	NEAB C	NEAB D	NEAB E
1 **Population**	1 **Population and settlement**	1 **Urban geography**	1 Weather and climate	1 The natural environment
2 Food and water*	2 **Agriculture**	2 **Economic geography**	2 Landforms and processes	2 **Population and settlement**
3 Energy resources*	3 Energy and natural resources*	3 People and the environment*	3 **Agriculture**	3 **Economic development and trade**
4 **The geography of manufacturing**	4 **Secondary industry**		4 **Industry and energy**	
5 **Urban geography**	5 **Tertiary industry**		5 **Economic development and trade**	
6 **Recreation, leisure and tourism**	6 **Transport and trade**		6 **Population and migration**	
7 Hostile physical environments			7 Settlement	

Table 17.2   Themes to be taught in the revised Geography
National Curriculum for England (human themes in bold)
(DfE, 1995)

Key Stage 1	Key Stage 2	Key Stage 3
Quality of environment	Rivers Weather **Settlement** Environmental change	Tectonic processes Geomorphological processes Weather and climate Ecosystems **Population** **Settlement** **Economic activities** **Development** Environmental issues

geography since the decline of regional geography during
the 1970s and its replacement by systematic studies. The
influential Schools Council Geography for the Young School
Leaver Project (Boardman, 1988) encouraged an emphasis
on human aspects of the subject and the relative neglect of
physical geography. However, the dominance of human
geography has been challenged in recent years from several
directions. Firstly, the University of London 16–19 A level
course has a people/environment focus and has experienced
a rapid growth in numbers taking it. Secondly, new GCSE
Criteria require 'balanced coverage of physical, human and
environmental aspects of the subject' (SCAA, 1995). Thirdly,
the new A/AS Subject Core requirements require 'an
integrative study' in each of three areas of geography:
environmental, physical and human (SCAA, 1993). Lastly,
the original Geography National Curriculum for England
and Wales increased the relative importance of physical
geography, environmental geography and the study of
places (DES, 1991; WO, 1991). In the revised National
Curriculum for Geography (DfE, 1995; WO, 1995) this shift
is evident in the proportion of human geography themes at
each key stage (Table 17.2). The supremacy of human
geography which has existed in schools since the mid-1970s
is unlikely to continue.

The second characteristic of human geography in schools
is the thematic approach which is used to structure the
subject. With a few exceptions, it is used by examination
syllabuses, the National Curriculum and textbook writers.

Table 17.3   An extract from A/AS Subject Core: geography (SCAA, 1993)

2.1   the core requires that all A and AS students gain in-depth knowledge, understanding and skills through the integrative study of:

(c)   a chosen human environment—its characteristics, processes (economic, social, political, cultural and physical), their interaction, consequent spatial outcomes and changes over time

Themes are taught differently by different teachers and different publishers, some encouraging an enquiry approach through the study of questions and issues related to the theme (Lambert, 1992) while others (Waugh, 1991) have used the themes more as a basis for transmitting particular concepts and information. In both approaches there has been a tendency towards fragmentation of human geography, as each theme is studied separately and in relation to a different place. This can obscure the way different aspects of human geography interrelate. Most examination syllabuses have, unwittingly, encouraged fragmentation by listing completely different examples for each theme, in an attempt to encourage the study of a variety of places across the world.

It is not until A level that official documents encourage more holistic studies within the domain of human geography (Table 17.3). The requirement for an integrative study of a 'chosen human environment' (SCAA, 1993) applies to all A level geography syllabuses for the examinations to be taken from 1996 onwards. An innovative syllabus which does not use the thematic approach to structure human geography is the A level syllabus of NEAB. It structures the human geography part of its syllabus according to different types of processes: economic, technological, social/demographic and political, thus emphasising geography as process rather than geography as content.

A third characteristic of human geography in schools is the emphasis on economic and social themes. On the economic side, mining, agriculture and manufacturing are studied more commonly than tertiary industry. On the social side, settlement patterns and functions are more commonly studied than the impact of unemployment. Cultural geography and political geography are relatively neglected. Themes such as electoral geography (Oxford and Cambridge Schools Examination Board) or the geography of war and peace (University of London, 16–19 syllabus) are included on new A level syllabuses, but as options.

Human geography in higher education

Human and regional geography take up about half of the modules available in most higher education institutions (HEIs) offering degrees in geography, although in some the proportion is considerably higher. Most degree courses in geography in England and Wales have one or two general introductory courses in human geography in the first year; some systematic modules emphasise the main subdisciplines of human geography, for example, social/economic/political geography, urban/rural geography, in the second year, although these terms are not always used in the module titles; and more specialised options, reflecting staff interests, are available in the final year. Students in most HEIs are able to specialise in human and regional geography to a greater or lesser extent in the second and particularly the final years of their degree courses. The converse is, of course, also true; some students take no modules in human and regional geography after the first or second year of their courses.

Since the mid-1970s the nature of human geography taught in higher education has changed (e.g. Gregory and Walford, 1989; Thrift, 1992; Johnston, 1993; Healey, 1994). This reflects both changes in the real world and changes in the discipline itself. Among the former are the increasing internationalisation of the economy; the growing concern over environmental issues; the break-up of the Soviet bloc; the debt crisis, particularly among Third World countries; the information revolution; and the growing awareness of differences between people, whether based on nationality, gender, religious or ethnic identity. Within geography there has been a parallel trend for an increase in the number of subdisciplines, as evidenced, for example, in the growth of study groups within the Institute of British Geographers (IBG). In 1994 there were 19 study groups, 12 of which were primarily concerned with human geography, compared with 8 groups with a human emphasis in 1974. In under-graduate courses there has been a significant growth in interest over this period in the study of social, cultural and political geography. These trends are likely to have a significant influence on school syllabuses in the next few years.

As a counter to the fragmentation of human geography, a number of geographers have emphasised in their writings the need to examine the interaction between the social, cultural, political and economic realms (e.g. Massey, 1984; Cooke, 1989) and this has been reflected in some courses (e.g. Crang, 1994).

Approaches to studying human geography

Approaches in schools

Although there is a variety of approaches to studying human geography in schools the legacy of the changes which took place in school geography in the 1970s remains; the positivist, scientific approach is still strongly represented in syllabuses and textbooks. Models (e.g. urban models), spatial theories (e.g. Christaller) and quantification (e.g. indices of accessibility) are commonly taught. Furthermore, school human geography is still very much concerned with general laws about people's behaviour. Many syllabuses and textbooks divide each theme into a series of key ideas or generalisations. In carrying out individual or group coursework, pupils often test hypotheses, use surveys to produce quantitative data, and produce generalisations in their study of human geography. The positivist approach is encouraged by some syllabuses more than others (e.g. NEAB syllabus B at GCSE) and by some authors more than others. The widespread use of textbooks by Waugh (1990a, b), even for syllabuses which encourage more varied approaches, helps to retain the positivist approach. The introduction of new themes, incorporating behavioural and welfare geography, e.g. the study of shoppers' preferences or the study of inequalities, has not really shifted the prevalent scientific methodology.

Since the introduction of GCSE syllabuses in 1986, the study of values and attitudes has been incorporated into all syllabuses to conform with national criteria. Some A level syllabuses and many lower school courses also include values education. In theory, this encourages the development of a more humanistic approach. In practice, however, different viewpoints are often considered as pieces of scientific evidence to be analysed, rather than being used to explore different meanings. Most school geography texts present data of all kinds as objective evidence to be accepted by pupils, rather than as something constructed and selected by people with different perceptions of the world. Many classroom activities demand reproduction of textbook information, analysis and generalisation rather than an interpretation of its meaning. Humanistic geography has made little headway in schools.

Much school geography is taught in an apolitical context in which issues of power and control are neglected. Often different viewpoints are discussed in terms of 'rights and wrongs' (Huckle, 1985, p. 187) rather than in terms of the power of different groups to influence events. The role of government is played down. Only a minority of texts (e.g. Huckle, 1988) and a minority of syllabuses (e.g. NEAB A level, and University of London, 16–19 A level) give due attention to political processes. The structural basis of human geography is rarely studied or investigated in pupil enquiries.

The original Geography National Curriculum discouraged humanistic and structural approaches to human geography. Almost all the statements of attainment referring to attitudes and values and to government policy were deleted from the the proposals for the Geography National Curriculum published in the National Curriculum Council consultation report (NCC, 1990). The revised 1995 Geography National Curriculum presents geography as an objective study, mostly set within an apolitical context. Teachers can, of course, interpret the new broad guidelines through alternative paradigms. Although the 1991 version of the Geography National Curriculum sets out a common curriculum, it has been variously interpreted by different teachers (Roberts, 1995) and different textbook authors. The most widely used books in the series, *Key Geography* (Waugh, 1991), introduce year 7 pupils to settlement, for example, through work on urban models and urban hierarchies. A less widely used series, *Enquiry Geography* (Greasley et al., 1991) introduces settlement through people's views of change in a village, suggesting a more humanistic approach. Relative sales of textbooks have an effect on approaches to human geography and clearly reflect in part the preferred approaches of teachers.

Approaches in higher education

Just as a growing subdivision has characterised the content

of human geography, a plurality of approaches to studying the discipline has emerged. The positivist viewpoints which dominate the study of geography in schools continue in higher education, but they have been challenged by humanist, structuralist, realist, feminist and postmodern approaches (Cloke et al., 1991). Jackson, Chapter 6 in this volume, outlines how these approaches reflect not so much linear trends in which new views gradually replace older ones, but rather multiple approaches which are actively contested. The complexity of some of these approaches means that their transfer to school syllabuses is unlikely to be undertaken as readily as some of the changes in the content of the discipline. It seems inevitable, however, that aspects of these different approaches will start to appear implicitly in the kinds of questions asked and in the interpretations given to the analyses of the results.

A greater variety of methodologies is also likely to spread into the study of geography in schools. The dominance of quantitative and extensive methodologies currently found in schools is likely to be supplemented by qualitative and intensive methods (Sayer, 1992). Ethnographic methods, for example, are playing an increasing role in the study of human geography in institutions of higher education.

The study of place and places

Place and places in schools

Since the decline in the study of regional geography in schools from 1970 onwards, places have been studied mostly in relation to themes studied in physical, human and environmental geography. Some syllabuses (e.g. NEAB syllabus B at GCSE and NEAB A level) have prescribed the specific places to be studied. A more common approach in GCSE and A level syllabuses, however, has been for teachers to be free to 'select their own examples to illustrate the key ideas' (NEAB, GCSE syllabus D). There has usually been a requirement to study places of different types at a range of scales and this is continued in the revised GCSE Criteria governing syllabuses to be taught from September 1996. The new A/AS level Subject Core also refers to 'a balanced sampling' of places. Some syllabuses list against a theme or a key idea whether it should be studied at a local, regional, national, international or global scale. This type of framework, used at GCSE and A level, has influenced what has happened below the age of 14. The result is that the most common approach to the study of places has been through case studies to illustrate whatever theme is being studied.

The advantage of this framework planning approach is that pupils study places at different scales and they study a variety of places all over the world. The variety can motivate and stimulate interest. It can increase general knowledge of the world, which a focus on a few places could not. Where there is not prescription there is also scope for teachers' expertise and local connections to add value to what is studied.

However, there are some disadvantages in the approach. There is a tendency for each case study to be at a particular scale, e.g. the small-scale study of a shanty town or the regional scale study of a rain-forest. It is like looking at the world through a series of fixed lenses, the shanty town through the telephoto lens and the Amazon Basin through the wide angle lens. This can make it difficult to make connections or, for some pupils, even to make distinctions between the various scales of study and to see how small-scale areas are affected by regional, national and international processes. What is needed for real understanding is the equivalent of a zoom lens which can change the focus from the small, local scale to the regional and back again quickly.

Another disadvantage of the case study approach is that, in an attempt to cover as much of the world as possible, particular places are studied for particular themes. India might be studied only for population, Peru only for shanty towns, California only for hi-tech industry. The danger of presenting pupils with misleading stereotypes is clear. A further disadvantage of the case study approach is that each place is studied for one theme and the processes which affect it. The interplay between political, social, cultural and economic processes can be lost.

The Geography National Curriculum has reinstated the study of places in their own right and not just as illustrative examples. The prescription of the study of particular countries in the original Statutory Orders has been replaced, from September 1995, by teacher choice in selection of localities and countries. One welcome feature of the revised Geography National Curriculum requirements for the study of place is the encouragement of the zoom lens approach mentioned above. In studying localities children have to be taught that they 'are set within a broader geographical context'. In studying countries pupils have to be taught about their 'interdependence with other countries'. When they study themes pupils have to study them 'within the context of actual places'. These wordings would seem to encourage a more holistic, meaningful study of places in schools. However, the list of what has to be studied on localities and countries (e.g. 'the main physical and human features' and 'the characteristics of two regions and their similarities and their differences') could encourage the collection of disparate facts about localities and countries rather than a meaningful study.

The study of places in the Geography National Curriculum provides a challenge for schools. What can they learn from the study of place and places in geography at higher education?

Place and places in higher education

Although in the 1970s and early 1980s regional geography never actually disappeared from higher education, it tended to play a subsidiary role to the systematic branches of the discipline. A new regional geography began to emerge, however, in the second half of the 1980s, which emphasised not only the uniqueness of places, but also their interdependence (Johnston, 1991). For example, Johnston (1984, p. 444) defined geography as being 'about local variability within a general context', while Massey (1984) argued that the character of places could be understood in terms of a series of layers of investment and disinvestment over time, each of which was influenced by the nature of the previous layers and the interaction with other places at that particular point in time. This approach showed that not only could the specificity of places be described, as the old regional geography had always done well, but that the character of places could be explained within a consistent theoretical framework.

Whereas the new regional geography has tended to concentrate on examining localities, albeit within a global context, the study of different parts of the world has continued, although sometimes within an interdisciplinary area studies context. The unit of analysis has, however, often shifted from that of individual countries to that of trading blocs, such as the European Union.

Pedagogy

Teaching and learning in schools

The National Curriculum defines what pupils should be taught in geography between the ages of 5 and 14 and what should be assessed, but it does not seek to prescribe *how* the subject is taught. There is a great variety in classroom practice at present, but there is encouragement from examination syllabuses, from the revised Geography National Curriculum, from some textbooks and from the legacy of Schools Council geography projects towards certain approaches to learning geography in schools. The most common recommendations include the adoption of an enquiry approach, the use of a wide range of resources, and the active involvement of pupils. The revised Geography National Curriculum includes statements in its level descriptions which encourage increasing involvement of pupils in the process of geographical enquiry. One approach to enquiry, 'the route to geographical enquiry' was defined by the 16–19 Geography Project (Naish et al., 1987). It provided two series of structured questions, to enquire into 'more objective data' and 'more subjective data', which could be adapted for the study of any issue. Geography was seen as being constructed by answering questions rather than consisting of the transmission of an already constructed geography. In theory, in this approach pupils have a part in devising the questions, in deciding on data to be used, and in interpreting these data. The examination papers for the University of London, 16–19 A level syllabus, for example, encouraged active involvement of pupils in an enquiry approach in both the examination papers it set—a decision-making examination paper and a paper which included a variety of data for interpretation—and both styles of question are retained in the new 16–19 syllabus assessment scheme (from 1997).

The testing of candidates' ability to interpret data is a common practice in GCSE and A level examinations. No GCSE or A level examination is based solely on regurgitation of knowledge. In spite of this, some pupils are prepared for these examinations inappropriately by dictated or copied notes. Many others learn geography through completing worksheets and written assignments. Other school geography departments encourage active participation through small group and whole class discussion, through role play activities and through the use of games and simulations. Whatever classroom style is used in school, all pupils studying geography will carry out some fieldwork and many will use computers.

Although there is a great variety of practice, school geography is usually learned in groups of 30 or fewer, it involves the study of varied resources, and most lessons involve interaction between teacher and pupils. Many lessons also include interaction between pupil and pupil. For the most part in both traditional and progressive departments there is a high degree of control by the teacher of the content of what is taught, the resources which are used and the activities in which pupils engage. The pedagogies of school contrast with the lectures, the seminar presentations, and the responsibility involved in independent learning during higher education.

Teaching and learning in higher education

Human geography is a popular discipline with students. This has meant that, when taken together with the growth in student numbers, the learning environment students experience on entering higher education differs significantly from that many have experienced in schools, further education colleges and sixth-form colleges. Many students studying A level geography do so in classes of no more than 10, 20 or 30 students , dependent on the type of institution they attend. In moving to higher education they often find themselves in first-year classes with 150–250 other students. Inevitably, with these numbers of students, lectures tend to dominate as a method for imparting knowledge. However, geographers have been innovative in finding ways of giving students a satisfying and deeper learning experience even when dealing with large classes; methods used include group work exercises, interactive lectures, and study packs (Jenkins et al., 1993, 1994). Increased reliance on textbooks for introductory classes seems likely in the future (e.g. Healey and Ilbery, 1993). Much emphasis in human geography classes is placed on discussion. This is an area in which many students coming straight from school lack confidence and skill.

As pressures on the unit of resource in higher education continue, a debate is beginning to occur on whether a national curriculum for higher education is desirable.

Although this seems unlikely in the immediate future, the sharing of resources between institutions is beginning to happen with, for example, the UK Teaching and Learning Technology Project based at the University of Leicester.

Changing human geography: the implications for schools and higher education

The contrasts outlined above between human geography as studied in schools and human geography as studied in higher education raise interesting questions about what would be the most desirable human geography curriculum in schools and in higher education.

Firstly, there are questions about the content of human geography. Some of the themes and issues studied in higher education might be thought to be inappropriate at school level. They could be thought, for example, to be too adult, too political, too complex, or unrelated to pupils' interests. This is worth exploring. Many pupils might be more interested in studying cultural and political geography, rather than what they are studying at present. Might pupils, as they gain independence to explore their local areas, be interested in learning about the gendered use of space and relating it to their own experiences? Might young people, with their own varieties of cultures and subcultures, enjoy studying places from particular cultural perspectives? Might young people enjoy discussing the power relationships between different groups involved in local and national decision-making? Why are some things included in the content of school human geography and others excluded? What are the criteria for selection? How are these debated and by whom? As power to decide content is increasingly centralised at the School Curriculum and Assessment Authority (SCAA), in GCSE groups and in A level boards and correspondingly removed from the professional judgement of teachers, it is vital that the debate takes place, and that it is informed by what is happening in higher education. There is a danger that centralisation, together with a widespread desire for a period of stability, will result in the content of school human geography becoming fossilised. School human geography needs to take account of changes in higher education, changes in the world and changes in the lives and needs of young people.

There are also implications for the content of human geography in higher education arising from the changes occurring in schools. For example, if central guidelines exist at 5–14, GCSE and A level, it needs to be asked whether there should be a centralised curriculum framework for first-year undergraduates. As the experience of human and regional geography of many students up to 18 becomes similar, there is the potential for introducing a more thought-provoking and different first year in higher education. For example, some HEIs base their first-year human geography course around the social, economic, political and cultural changes occurring at a global scale and the way their impact on different areas is modified by local conditions.

Secondly, there are questions about the approach to teaching geography. There is scope for extending the way geography is studied in school to explore meanings and interpretations of places and to include pupils' meanings as part of this study. Teachers could learn from the successful use of humanistic fieldwork methods in higher education. Ellis (1993) provides a good example of such work with first-year undergraduates in Norway. There is also scope for incorporating more political, social and economic theory at A level, so that geography is understood in relation to the underlying structures of our society.

One of the problems associated with the introduction of new approaches to human geography and the study of place in schools is the provision of suitable textbooks. Many of the texts used in higher education seem difficult to teachers as they are based on an understanding of aspects of social and political theory which teachers have not studied. There is a need for people to be translators or mediators between the two worlds. In theory, professionals working on either side of the higher education/school divide could do this, but they are constrained by the structures within which they work. University lecturers are under pressure to publish. They have a vested interest in spending their time writing academic books and articles which are regarded more favourably than contributions to school textbooks in terms of university research ratings and promotion. Teachers have limited time to write and there are few opportunities for periods of secondment and study. Some teachers, when given a chance to study, prefer to study for MA and PhD degrees, either for personal interest or to satisfy career aspirations, rather than write school textbooks.

There are also questions to be resolved about the teaching of place in schools. At present the new demands to teach place as part of the National Curriculum have not been accompanied by new approaches to understanding the meaning of place.

Within higher education, perhaps the greatest challenge facing departments of geography in their approach to teaching the subject is how to respond to the students arriving in their institutions who, from the age of 5, have predominantly been exposed to a system of learning in interactive classroom situations. Many pupils will have been encouraged to become actively engaged in their learning through enquiry approaches and resource-based learning. Higher education institutions have much to learn from schools, not least the need to shift a greater proportion of their efforts from teaching through lectures to facilitating learning through more student-centred methods. Many geography departments have already begun to move in this direction, but more consideration needs to be given to the relative effectiveness and efficiency of the different ways in which students learn.

All these questions provide a challenge for schools and higher education. In facing up to these challenges the links between geography in schools and in higher education need to be emphasised. However, the distinctive 'missions' of the two sectors may lead them in different directions in their approaches to studying human and regional geography. Whereas in schools national changes are encouraging a greater similarity in the kind of human and regional geography that students experience, it is as yet unclear whether the changes occurring in higher education will also lead to a greater similarity or diversity within that sector.

References

Boardman, D. (1988) The impact of a curriculum project: geography for the young school leaver. *Education Review*, Occasional Publication no. 14. The University of Birmingham.

Cloke P., Philo C. and Sadler, D. (1991) *Approaching Human Geography*. PCP, London.

Cooke, P. (1989) *Localities: the Changing Face of Urban Britain*.

Unwin Hyman, London.

Crang, P. (1994) Teaching economic geography: some thoughts on curriculum implication. *Journal of Geography in Higher Education*, **18**, 106–113.

DES (Department of Education and Science) (1991) *Geography in the National Curriculum*. HMSO, London.

DfE (Department for Education) (1995) *Geography in the National Curriculum: England*. HMSO, London.

Ellis, B. (1993) Introducing humanistic geography through fieldwork. Journal of Geography in Higher Education, **17**(2), 131–139.

Greasley, B., Ranger, G., Williamson, E. and Winter, C. (1991) *Enquiry Geography*. Hodder and Stoughton, London.

Gregory, D. and Walford, R. (eds) (1989) *Horizons in Human Geography*. Macmillan, Basingstoke.

Healey, M. (1994) Teaching economic geography in UK higher education institutions. *Journal of Geography in Higher Education*, **18**, 70.

Healey, M. and Ilbery, B. (1993) Teaching a course around a textbook. *Journal of Geography in Higher Education*, **17**, 123–129.

Huckle, J. (1985) Values education through geography: a radical critique. In D. Boardman (ed.) *New Directions in Geographical Education*. Falmer Press, Lewes, pp. 187–197.

Huckle, J. (1988) *What We Consume, the Teachers' Handbook*. World Wildlife Fund, Richmond Publishing, Godalming, Surrey.

Jenkins, A. et al. (1993) Teaching large classes in geography: some practical suggestions. *Journal of Geography in Higher Education*, **17**, 149–162.

Jenkins, A. et al. (1994) Control and independence strategies for large geography classes. *Journal of Geography in Higher Education*, **18**, 245–262.

Johnston, R.J. (1984) The world is our oyster. *Transactions, Institute of British Geographers*, **9**, 443–459.

Johnston, R.J. (1991) *A Question of Place: Exploring the Practice of Human Geography*. Blackwell, Oxford.

Johnston, R.J. (ed.) (1993) *The Challenge for Geography; a Changing World; a Changing Discipline*. Blackwell, Oxford.

Lambert, D. (ed.) (1992) *Cambridge Geography Project: Jigsaw Pieces*. Cambridge University Press, Cambridge.

Massey, D. (1984) *Spatial Divisions of Labour*. Macmillan, Basingstoke.

Naish, M., Rawling, E. and Hart, C. (1987) *The Contribution of a Curriculum Project to 16–19 Education*. Longman, Harlow.

NCC (National Curriculum Council) (1990) *Geography 5–16 in the National Curriculum: a Report to the Secretary of State for Education and Science on the Statutory Consultation for Attainment Targets and Programmes of Study in Geography*. NCC, York.

Roberts, M. (1995) Interpretations of the geography national curriculum: a common curriculum for all? *Journal of Curriculum Studies*, **27**(2), 187–205.

Sayer, A. (1992) *Method in Social Science*. Hutchinson, London.

SCAA (School Curriculum and Assessment Authority) (1993) *GCE Advanced and Advanced Supplementary examinations. Subject Core for Geography*. SCAA, London.

SCAA (1995) *GCSE Criteria for Geography*. SCAA, London.

Thrift, N. (1992) Apocalypse soon, or, why human geography is worth doing. In A. Rogers, H. Vile and A. Goudie (eds) *The Student's Companion to Geography*. Blackwell, Oxford, pp 8–12.

Waugh, D. (1990a) *The British Isles*. Nelson, Walton-on-Thames.

Waugh, D. (1990b) *Geography, an Integrated Approach*. Nelson, Walton-on-Thames.

Waugh, D. (1991) *Key Geography: Foundations*. Stanley Thornes, Cheltenham.

WO (Welsh Office) (1991) *Geography in the National Curriculum (Wales)*. HMSO, Cardiff.

WO (1995) *Geography in the National Curriculum (Wales)*. HMSO, Cardiff.

The experience of physical geography in schools and higher education

18

John Davidson and Derek Mottershead

The experience of physical geography

What do adults remember of their experience of geography at school? For those who have only studied geography as far as age 14 or 16, the answer is often their *physical geography*. They may remember their fieldwork in a local river valley or on a trip to the coast, or collecting weather data, or studying the effects of a volcano or hurricane. These topics appeal to pupils and can often be taught in an imaginative and innovative way. For students who have studied the subject at A level and beyond, it is often still the physical geography that is most memorable. Former A level students will return to their school or college and describe enthusiastically travel experiences where they saw the landforms and landscapes that had featured in their geography lessons. It is this same interest in landscapes and landforms that is frequently cited as a reason for a student going into teacher training with a geography subject specialism in mind.

Now a wealth of images in books, on television and on CD-ROM brings the physical world of landscapes, landforms and climatic regions into the home (see, for example, Goudie and Gardner, 1992), and many seek to interpret and understand this world rather than just view it. Physical geography for children and adults alike is exciting, stimulating, memorable and popular. The high ratings of recent

Geography into the Twenty-first Century, Edited by E.M. Rawling and R.A. Daugherty,
© 1996 John Wiley & Sons Ltd.

television programmes on *Antarctica* and *Great Journeys* has not just been due to the pictures of wildlife. The 7-year-old expressing wonder at the prospect of rising sea-levels, and the third-year undergraduate student investigating coastal sediment cells are both attracted to the idea of trying to understand, at different levels, the workings of the physical environment. Furthermore, there is a growing general interest in unravelling how the physical world works to help our management of physical systems. The experience of physical geography available to pupils throughout school and college is potentially, therefore, one of both fascination and relevance.

The basis for a sound physical geography in schools is now clearly set out. The revised National Curriculum Order for Geography (DfE, 1995; WO, 1995), the GCSE National Criteria for Geography (SCAA, 1995) and the A/AS Subject Core (SCAA, 1993) all ensure that both physical geography and the complex relationships of people with the physical environment feature strongly in courses, schemes of work and assessment. In higher education, physical geography has traditionally been strong and maintains an important position. However, the many changes currently taking place in education at all levels mean that there are no grounds for complacency. The experience of exciting and relevant physical geography can only be ensured if effective teaching with appropriate resources can occur, meeting the needs of those in schools and universities. This chapter examines the experience of physical geography in primary schools, secondary schools and higher education and how this fits the needs of pupils and students. Where appropriate, issues are identified which need to be addressed if physical geography is to continue to thrive. The conclusion makes suggestions for future action and indicates which organisations might take up these initiatives.

Physical geography 5–14

Children in primary schools need a basic grasp of physical geography to understand their own environment, and to make sense of the places they are studying. In secondary schools, there is a need to build on these foundations by extending the range and scale of physical geography topics and the depth of study.

The physical geography of places has always been taught in good primary schools, although often through general topics rather than under a heading of geography. In the HMI publication, *Geography from 5 to 16* (DES, 1986), opportunities were highlighted for primary schoolchildren to develop a more rigorous base in physical geography through the study of weather, local landscape features, rocks, soils and biogeography and, particularly, through developing appropriate skills and an enquiry approach. The original National Curriculum for England and Wales (DES, 1991; WO, 1991) required at Key Stages 1 and 2 some basic knowledge and understanding of rocks, soils, the water cycle, rivers and river landforms and weather measurement and patterns. In secondary schools, although physical geography was a component of most 14–16 GCSE courses, there was some concern that it was being covered less and less in lower secondary school schemes of work. In 1987, Stoddart (1987, p. 331) writing for an academic journal, alerted geographers to the dangers of neglecting the physical basis of the subject:

> Geographers have forgotten—it is extraordinary to say so—that some parts of the Earth are high, others low; some wet, others dry; some deserts, others covered by forest and grassland and ice. No one seems to mention these days that two-thirds of the surface of the planet is covered by the sea. These are elemental categories of human existence in which geography must deal.

Certainly many school syllabuses below O level/GCSE in the mid-1980s included little basic physical geography or weather, and occasionally none at all. The introduction of the original Geography National Curriculum partly reversed this trend, with the inclusion of statements of attainment on a wide range of aspects of physical geography drawn from weather and climate, soils and vegetation, and landscape processes and features. However, while the early 1990s were marked by secondary schools redesigning their lower school geography syllabuses to include these basic features, the practical experience in many schools became one of a content-laden course which was difficult to teach effectively. Geography started to become fragmented as teachers attempted to tick off each statement of attainment. Some textbooks reinforced this trend.

The revised National Curriculum (DfE, 1995; WO, 1995)

reduces substantially the total amount of physical geography to be covered by the end of Key Stage 3, but clarifies the focus and keeps the important elements, providing opportunities for progression and enhancement. At Key Stage 1, pupils are to be taught to describe physical features of places, and to begin to look at weather and how it affects us. At Key Stage 2, these skills are extended and specific thematic studies address rivers, their processes and their effects on the landscape, and weather conditions and patterns both in the home region and around the world. Within a few years all pupils at age 11 should be able to describe and compare landscapes, measure physical and climatic conditions and understand the basic work of rivers. This should, with the development of associated skills of data collection, analysis and enquiry, form a sound basis for further progression in physical geography in secondary schools.

At Key Stage 3, although the study of rocks is firmly placed in science, geography teachers are required to cover four major physical geography themes, plus a series of enquiry and fieldwork skills, many of which can be developed through physical geography. The themes are:

1. Tectonic processes, with requirements including the global distribution of earthquakes and volcanoes; a study of the nature, causes and effects of one or the other; human responses to the hazards.
2. Geomorphological processes, with requirements including the development of landforms associated with rivers or coasts; the causes and consequences of river floods, coastal erosion or coastal flooding; the human responses to the hazards posed.
3. Weather and climate, with requirements including the difference between weather and climate; the water cycle; variations in conditions from place to place.
4. Ecosystems, with requirements including the relationship between climate, soils and vegetation; a study in detail of at least one vegetation type.

The new National Curriculum Order seems to provide a sound foundation in physical geography for all pupils 5–14. However, the challenge arising from this is twofold. Firstly, how to ensure that the minimum national guidelines are developed into good physical geography, especially in the many primary schools where there is little geographical experience, and where the physical aspects of geography may have been neglected pre-National Curriculum. And secondly, how can the subject be made exciting and stimulating for pupils of all ages?

There are a number of concerns about the current situation. Firstly, many primary teachers have received little subject-specific training in geography and their knowledge base of physical geography, in particular, may be limited. Secondly, in some primary schools the teaching of physical geography as a discrete topic, rather than as part of a wider study of places, may result in children encountering too much detail or inappropriate language. Thirdly, the progression from Key Stage 2 to Key Stage 3 is sound in theory, but a much greater dialogue between primary and secondary schools will be needed in some areas to avoid overlap and duplication, particularly over the choice of places studied and the use of resources. Secondary schoolteachers may underestimate how far a child aged 9 or 10 has already developed an awareness of physical geography, which could be developed and enhanced at Key Stage 3. For the subject to be enriched and to build on the Key Stage 1 and 2 foundations, it is important that Key Stage 3 teachers themselves all have a strong grounding in the subject. There is already a danger that some non-specialist teachers will cover only the minimum requirements relying possibly on dated or inappropriate resources. Finally at Key Stage 3, geography teachers will also need to develop links with their colleagues in science in order to build on the common areas such as properties of rocks and weathering processes.

These concerns are already being addressed in a number of ways. In some local education authorities (LEAs), secondary and primary teachers of geography find opportunities to work together and such co-operation and co-ordination will help address the problem of Key Stage 2 and 3 progression. Curriculum guidance such as the ideas about how to build physical geography into a wider study of places, as suggested by Krause et al. (1993) is helpful, and the publications of the Geographical Association (GA) and its journal *Primary Geographer*, also provide a range of targeted advice and practical suggestions. In-service courses for primary teachers run under the grants for education support and training (GEST) programme and through the GA's conferences and summer schools have included physical geography, and at local levels in-school training led by the geography co-ordinator can be successful in raising the expertise of the staff as a whole.

For Key Stage 3 teachers, the in-service training courses provided by many LEAs already deal with the integration of physical geography topics with study of places and development of skills. Some schools have also developed combined fieldwork days with their science colleagues looking at both rocks and landforms/landscapes, while others have developed enquiry approaches in geography (for example investigating building stones and quarry locations within an urban area) which build on the knowledge of rocks gained through science and encourage applications of this knowledge. There is a need for such activities to be more widely spread across England and Wales, and to ensure that appropriate fieldwork opportunities, particularly at local level, are recognised and used by schools.

For physical geography to be exciting and stimulating at Key Stages 1, 2 and 3, teachers need topical, visual and interactive materials and the confidence to make use of a range of enquiry-based teaching and learning approaches. At Key Stages 1 and 2, relevant resources are now becoming available. For instance, some excellent video material has recently been produced with suitable commentaries for young children. However, limited finance may restrict the use which can be made of photosets, videos or CD-ROMs, or prevent some schools (both primary and secondary) from acquiring equipment such as weather-recording devices. Another concern is the relatively narrow range of teaching materials that exist at present for the teaching of physical geography at primary level. The GA's *Focus on Castries, St Lucia Photopack* (Bunce et al., 1992) and *Images of Earth: a Teachers' Guide to Remote Sensing in Geography at Key Stage 2* (Barnett et al., 1995) are successful examples of what is already published but many perceive a need for a wider range of affordable materials to be made available. There is also a wish to raise awareness of sources of materials that can be adapted at this level, such as, for example, the publications of the Royal Meteorological Society, and the Natural History Museum.

For Key Stage 3, there has traditionally been a wide range of textbooks and other resources available and this is now being extended particularly in the area of information technology (IT) and fieldwork equipment. *Geography and IT: Investigating Weather Data* (GA/NCET, 1995) is an example of how pupils can study aspects of weather using data-handling packages. The development of such interactive classroom materials is a welcome step; however, it is crucially important that this is followed up by sound in-service support on appropriate classroom approaches. The Geography/IT Support Project run by the GA and NCET is providing the beginning of such assistance, but it will need to be extended more widely at LEA and local level.

Over the next few years, there will doubtless be many excellent resources and teaching approaches developed to support physical geography at all key stages. They will only have a beneficial effect on school geography if they are disseminated widely, through targeted in-service activities, and by means of booklets and articles describing school experiences in appropriate publications.

Physical geography in the 14 to 19 age range

From the end of Key Stage 3 at age 14, geography becomes an optional subject within the school curriculum. Geography is currently a successful subject at GCSE with around 260000 candidates per year (see Appendix 3), and if the subject is to retain this position under the pressures of a post-Dearing world, it will need to maintain its character as a stimulating and relevant subject. GCSE provides opportunities for fieldwork in physical geography, and the pattern in many schools will be to link this to a local investigation to develop enquiry skills, by studying a section of river, a coastline or an area of rural land. The revised GCSE courses from 1996 should also provide a common foundation in physical geography, since the GCSE Criteria build on National Curriculum foundations. Prescribed topics include tectonic processes, geomorphological processes, weather and climate, and ecosystems. Study of these should be undertaken in the context of places at different scales and should include the effects of human activity. However, some topics such as glaciation, traditionally covered by schools at GCSE, may receive less attention. Overall, pupils following the new GCSE courses should receive a firm foundation in the basic principles of physical geography, although their knowledge may be limited to a narrow range of examples. There is the potential for some examination boards to give greater emphasis to physical geography in their syllabuses and this should enable teachers to make greater use of a range of textbooks and other materials that are available at this level. However, guidance on appropriate practical investigations, that can take place without leaving the school grounds, will be important where there are constraints on fieldwork.

At A level, physical geography has suffered from a decline in emphasis in some syllabuses over recent years. Some A level boards have focused more on the applied aspects to the exclusion of a systematic grounding in basic principles, and it has been possible with some syllabuses to avoid covering major areas of physical geography and still attain a high grade in the subject. The agreed common core at A level which is the foundation of the new syllabuses which have been produced during 1994–95, ensures the place of the basic principles of physical geography. The requirements include, among other things:

1. A theme which emphasises the interaction between people and their environment at different spatial scales, and focuses on the relevant systems and processes, their outcomes, changes through time and consequent issues, responses and strategies.
2. A chosen physical environment—its characteristics, processes (terrestrial, atmospheric, biotic and human), their interaction, consequent spatial outcomes and changes over time.

As a result, from 1997 (first examination for the new syllabuses) all students should have a grounding in some aspects of physical geography. More importantly, students will have to study aspects of the atmosphere, biosphere, lithosphere and hydrosphere to cover this requirement fully, and inclusion of the dimensions of time, magnitude and rate of change may be particularly welcomed. Schools are also being encouraged, through the core and the new syllabuses, to equip their pupils with investigative skills. Overall, the common core should enhance the role of physical geography at A and AS level.

However, changes at A level, and to a lesser extent at GCSE, are also a major challenge for the many schools which have previously followed syllabuses with a limited coverage of physical geography. There are four main concerns. First of all, A level and GCSE teaching is not always the realm of specialist subject practitioners. Non-specialist teachers or geography graduates whose degree and A level background did not include much physical geography may find difficulties delivering parts of the new syllabuses, especially at A level. There is a possibility that such teachers could end up covering the bare minimum of physical geography unless they can be given appropriate encouragement and support. There is a clear need for up-to-date, practical, student-centred materials in physical geography, especially at A level, with appropriate teacher guidance plus in-service training where possible.

Secondly, at A and AS levels, the ability base of those starting courses is widening. This creates problems as the readability levels of the physical geography sections in a number of A level texts make them inaccessible to some pupils. Approaches to teaching may need to become more practical and pupil centred. Many opportunities exist in A level physical geography for using interactive, data-based resources and good topical case studies. Indeed, the requirement to study physical processes in a specific place is a fundamental part of the new core. However, there is a limited range of such materials at present. In the short term, articles in journals such as *Geography* and *Teaching Geography* could help, while in the longer term there may well be possibilities for schools to access such information through E-mail links or through CD-ROMs, the latter providing potentially a topical and, in theory, an affordable way to convey such information effectively.

A third concern is that since much of the best physical geography is taught, at least partly, through fieldwork, the financial costs and time constraints on fieldwork may prove prohibitive, A solution is to bring physical geography into the school grounds and classroom. The use of IT resources such as CD-ROMs and data-logging equipment within the school, of simulation models such as wave or river tanks, and of air photo and satellite image packs all provide ways forward for the A and AS student. There is also considerable scope for laboratory experimentation in geography, and for the use of data logging within the immediate school area. In all these activities, curriculum materials and guidance in appropriate use will be needed.

A final concern is how far National Curriculum science provides a foundation for coping with the science within A level physical geography. Teachers of geography will need to be aware of what has been covered in the science curriculum, and may have to continue to adapt their teaching to make physical geography accessible to those with a limited scientific background.

The GA and the British Geomorphological Research Group (BGRG) could play a prominent role in facilitating the development of materials, and publications to address some of the needs of those delivering the new A level and GCSE syllabuses. There is also scope for in-service courses at national and local level covering physical geography case studies and appropriate practical techniques. The GA already provides some of these, but there is potential to develop them more widely at regional level. Guidance will also be needed on obtaining and evaluating resources like CD-ROMs and remotely sensed databases, if these resources are to be cost effective.

Physical geography in higher education

Physical geography has a clear position in higher education geography programmes, and the foundations of the subject are strong (see Gardner, p. 95; Gregory, 1992; Goudie, 1994). It is also highly relevant to the world community, although it is apparent that issues of physical geography are not always fully understood by some of the major environmental decision-makers of society.

Most major higher education geography programmes embrace foundation courses in physical geography in year 1, with a range of core courses, some perhaps optional, in year 2, and a variety of more specialist options in year 3. The student may find him/herself in a class of 100–300 students in foundation courses, with progressively smaller classes with increasing optionality and specialisation through years 2 and 3. The newly inducted student in higher education will be faced with the experience of a foundation course the nature of which tends to be a broad-based perspective across the field of physical geography, commonly embracing geomorphology, biogeography and climatology, and often with an emphasis on physical geography processes. The flavour of such courses can be gleaned from the course texts which are commonly employed, such as Briggs and Smithson

(1985), Goudie (1993) and White et al. (1992).

The teaching style of the foundation course tends to be lecture based, although such a course may be supported by small group work in tutorials, practical classes and workshops. Although opportunities for small group and interactive work in physical geography may well be rather limited in the first year, the decreasing class size thereafter renders this kind of teaching more practicable. Fieldwork, which is a component of most higher education geography programmes and may take place at any level within the programme, offers particularly valuable opportunities for small group and practical work.

Students coming through from schools and colleges over the next few years may be expected, on the other hand, to be well versed in enquiry approaches and decision-making, as a consequence of more student-centred styles of teaching through primary and secondary education. They will expect discussion-based teaching rather than lectures. They will expect stimulating courses in which they have a clear part to play, and a chance to build on their existing knowledge base. They will have expectations of interactive, resource-based teaching and exciting approaches. In many cases they will be well versed in data handling and the use of IT. Higher education staffing levels and resource provision may not facilitate the interactive approaches which such students will expect. This will create both opportunities and issues for higher education. There will be an increased need to develop courses based on student-centred resources using IT and perhaps video technology.

In higher education in the mid-1990s, a greater number of students of a wider range of ability are already seeking to study geography, and these students exhibit significant differences in knowledge and understanding of physical geography from their predecessors. The choice of routes through the GCSE and A level syllabuses has resulted in some students displaying a partial knowledge of physical geography. In addition, the increasing trend to topic-based studies in some syllabuses has encouraged superficiality and a neglect of the fundamental processes of physical geography. Students with this background tend to lack an understanding of the systematic range of physical geography, and of the nature of the processes within it. Because of the way in which A/AS level subject choices have been made, they may also lack a grounding in the basic sciences of physics, chemistry and biology, and also geology, and may have had very limited exposure to laboratory analytical processes. Such students embarking on higher education geography programmes which have a substantial component of physical geography may then need some supplementary scientific background. The effect of this inadequate grounding is likely to persist until the late 1990s. After this, cohorts which have experienced the National Curriculum in its entirety *and* the A/AS Subject Core will come through into higher education. It remains to be seen how effectively these central guidelines for both geography and science will have laid the foundations for good physical geography.

Pointers for the future development of physical geography in schools and higher education

In schools

If schools are to produce enthusiastic students with a genuine interest in physical geography, equipped with a basic set of skills within the subject, and demonstrating an understanding of some of the major principles, then there are five major needs, detailed as follows.

Interpretation of curriculum documents

The new Geography Order, and the new GCSE and A/AS level syllabuses, will need to be supported by a range of guidance material and in-service provision, illustrating how to plan and implement good physical geography starting from the central guidelines. This is particularly the case when non-specialist teachers are entering less familiar territory.

Topical, relevant teaching materials at all levels, linked to the new technology where possible

There is a major need for economical, easily accessible case study material and teaching ideas which are up to date and relevant to the new courses.

The new technology, particularly CD-ROMs and Windows

data-handling packages, provide potential opportunities for student-centred learning in physical geography. New materials embracing IT have been published on water, (NCET, 1994) and on weather data (GA/NCET, 1995). There are some well-researched video programmes available, but there is also a wealth of images and data now appearing on CD-ROM, some of which lack supporting curriculum materials. Teachers urgently need the means to evaluate CD-ROMs before committing their limited budgets.

Inclusion of physical geography units in teacher training courses

In teacher training it will be important to ensure that some physical geography is provided during the training period. This will be especially significant over the next five years when a cohort of trainee teachers who have received little physical geography from their own school or university education, will be coming through the system. At the same time those being trained substantially in schools should receive sufficient guidance with their physical geography.

Appropriate INSET for teachers already in post

Their is a vital need to ensure that teachers in post, in both primary and secondary schools, receive opportunities to update their physical geography. All short in-service courses should have a component of physical geography. Significantly, this can often be combined with the development of skills. Regional courses on IT in geography, using physical geography examples, occurred in 1994/95 enabling teachers to develop IT expertise while learning how to enhance physical geography teaching (Geography/IT Support Project). Similar courses have been able to demonstrate physical geography themes while focusing on fieldwork skills. The regional events of the GA and the Royal Geographical Society (RGS) provide a major opportunity for physical geography in-service activity.

In higher education

There are substantial implications for higher education of the wide range in ability and motivation of students who may be expected to enrol in the future. These implications embrace the design, the mode of delivery and the resourcing of courses.

1. There may be a need to explain the relevance of physical geography to students whose prior study may not have sufficiently exposed them to its study. All committed physical geographers in higher education will have their own rationale for the study of physical geography, but should not overlook the need to explain this to students.

2. Despite the existence of the A/AS Subject Core, it may still be necessary to reconsider the content of the first-year foundation course in higher education and even to devise extra support for those students whose A level experience did not promote a thorough grounding in physical geography. In addition there will still be those non-standard entrants who have arrived via Access[1] and other non-traditional routes.

3. The need to engage the individual student in large classes suggests that higher education teachers will need to adopt a more student-centred approach than has hitherto been the norm. This may include, for instance, the development of resource-based activities and simulations, and the deployment of graduate teaching assistants to afford a greater degree of contact between student and teacher in the large class.

4. The need to teach practical skills to large classes in both field and laboratory carries implications for many aspects of resource provision. Careful consideration will have to be given to teaching space, laboratory facilities, numbers of supporting staff, appropriate equipment and suitable safety standards.

Conclusion

In the current climate of continuing change in the school curriculum, there is increasing need for close liaison between school geography and higher education, in order to ensure continuity in the geographical education of students (see Bradford, p. 277). The interface between school and higher education geography, in relation to physical geography, is probably under greater strain at present than at any other time in recent history. It is therefore imperative that those bodies with an informed interest in this interface, both subject-wide organisations such as the GA, and the Council of British Geography (COBRIG), and more specifically focused bodies such as the BGRG and the Earth Science Teachers' Association (ESTA), foster activities which address subject and student needs at this interface.

Specifically, some ways forward might include the following. ESTA could work more closely with the GA to develop resources that will benefit both geography and science. The GA through its working groups could interpret the various curriculum documents and show, through examples, how the teaching of physical geography can be enriched. The results can be made available through the association's three journals or through new publications. The GA can also maintain and enhance its existing role as regular provider of articles to update physical geography. Special attention could be given to these through *Primary Geographer*. BGRG could make case studies available to teachers of recent research in a form that could be used in class through publications or through further presentations at the GA conference following the excellent model of the annual conference in Oxford in 1994. Finally, COBRIG is in a powerful position to press for physical geography to be included in in-service education and training and to encourage appropriate courses to be set up and funded.

This is a time of great opportunity in physical geography, when, with a common purpose, a range of organisations and teachers in schools and in higher education can take the subject forward into the twenty-first century, as a strong entity and as a central part of geography.

[1] Access courses are pre-entry programmes for mature entrants into higher education.

References

Barnett, M., Kent, A. and Milton, M. (ed.) (1995) *Images of Earth: a Teacher's Guide to Remote Sensing in Geography at Key Stage 2*. GA, Sheffield.

Briggs, D.J. and Smithson, P.A. (1985) *Fundamentals of Physical Geography*. Hutchinson, London.

Bunce, V., Foley, J., Morgan, W. and Scoble, S. (1992) *Focus on

Castries, St Lucia photopack. Geographical Association, Sheffield.

DES (Department of Education and Science) (1986) *Geography from 5 to 16, Curriculum Matters 7.* HMSO, London.

DES (1991) *Geography in the National Curriculum (England).* HMSO, London.

DfE (Department for Education) (1993) *Geography in the National Curriculum (England).* HMSO, London.

GA/NCET (Geographical Association and National Council for Educational Technology) (1995) *Geography and IT: Investigating Weather Data.* GA, Sheffield and NCET, Coventry.

Goudie, A.S. (1993) *The Nature of the Environment* (3rd edn). Blackwell, Oxford.

Goudie, A.S. (1994) The nature of physical geography, a view from the dry lands. *Geography,* **79**(3), 194–209.

Goudie, A.S. and Gardner, R. (1992) *Discovering Landscapes in England and Wales.* Chapman and Hall, London.

Gregory, K. (1992) Changing physical environment and changing physical geography. *Geography,* **77**(4), 323–335.

Kent, A. (ed.) (1985) *Perspectives on a Changing Geography.* GA, Sheffield.

Krause, J., Campion, K. and Carter, R. (1993) Putting physical geography in the right place. *Primary Geographer,* **13**, 17–21.

NCET (National Council for Education Technology) (1994) *Water/Excel: Key Stages 3 and 4.* NCET, Coventry.

SCAA (School Curriculum and Assessment Authority) (1993) *GCE A and AS Subject Core for Geography.* SCAA, London.

SCAA (1995) *GCSE Subject Criteria for Geography.* SCAA, London.

Stoddart, D.R. (1987) To claim the high ground: geography for the end of the century. *Transactions of the Institute of British Geographers,* **12**(3), 327–336.

Summerfield, M. (1991) *Global Geomorphology.* Longman, Harlow.

White, I.D., Mottershead, D.N. and Harrison, S.J. (1992) *Environmental Systems, an Introductory Text* (2nd edn). Chapman and Hall, London.

WO (Welsh Office) (1991) *Geography in the National Curriculum (Wales).* HMSO, Cardiff.

WO (1995) *Geography in the National Curriculum (Wales).* HMSO, Cardiff.

Teaching environmental issues in schools and higher education

19

Graham Corney and Nick Middleton

Introduction

Environmental issues are concerns that have arisen as a result of human impact on the natural environment and the ways in which the natural environment affects human society. The study of environmental issues, by its very nature, requires and encourages the multidisciplinary approach that has long been a facet of geographical study. Although other disciplines have recently taken an interest in some environmental issues, study of the interactions between people and their environment has formed an increasingly important focus of school geography in the last half-century, and has always formed one of the paradigms or schools of thought in academic geography (see, e.g. Pattison, 1973; Holt-Jensen, 1980; Cloke et al., 1991). There is, however, no room for complacency about the role of geography in the study of environmental issues (see O'Riordan, Chapter 8).

In this chapter, we look at the place of environmental issues in geography in schools and higher education, the nature of the subject matter and the teaching approaches used. The first section is about the situation in primary and secondary schools, the second reviews the situation in higher education, and the conclusion summarises some implications of these developments for interactions between schools and higher education.

Geography into the Twenty-first Century, Edited by E.M. Rawling and R.A. Daugherty, © 1996 John Wiley & Sons Ltd.

Environmental issues in schools

A review of current practice

Before the introduction of a National Curriculum in the schools of England and Wales, teaching about environmental issues in geography varied between schools, depending on their particular curriculum. A summary of practice would have included the following features. In primary schools, environmental matters were strongly represented through 'local studies', characterised by an interdisciplinary approach involving fieldwork. In the early secondary years they were represented in specifically geographical local studies and in themes such as pollution, British National Parks, and Third World development. In GCSE and A level teaching, coverage of environmental issues tended to relate to examination board syllabuses, with the strongest emphasis given in the Midland Examining Group/Welsh Joint Education Committee (MEG/WJEC) Avery Hill syllabus at GCSE, and the University of London Geography 16–19 A level syllabus where they provide the unifying philosophy.

In the National Curriculum, 'Environmental geography' was specified as one of the five attainment targets in the English and Welsh Orders (DES, 1991; WO, 1991) and environmental issues could be identified in the other four. Its importance appears to have increased marginally in the new Orders for Key Stages 1–3 (DfE, 1995; WO, 1995). In addition, although the status of cross-curricular themes is unclear, the contribution of geography teachers to students' environmental understanding was supported in various documents, especially *Curriculum Guidance 7* (NCC, 1990), and the Curriculum Council for Wales Advisory Paper 17 (CCW, 1992).

At a basic level, then, the opportunity for geography teachers to contribute to their students' environmental education is acknowledged, but our real concern is with the nature of this contribution. Is it to be simply 'teaching about the environment', which we feel is insufficient on its own, or is there to be a more committed approach, relevant to the world's environmental predicament, through 'teaching for sustainability'?

Approaches to environmental education

Three approaches to environmental education are frequently recognised (see e.g. NCC, 1990, pp. 7–12; Sterling and Cooper, 1992, pp. 2–3):

1. *Education about the environment* promotes knowledge and understanding of physical and human systems and their interaction. This approach can be described as 'education for environmental management and control' (Huckle, 1993a, p. 61).
2. *Education in (or through) the environment* promotes skills and it may promote concern, using the environment as a resource for learning, through fieldwork which is frequently student-centred. This approach can be described as 'education for environmental awareness and interpretation' (Huckle, 1993a, p. 61).
3. *Education for the environment* builds on the previous two approaches to promote informed concern, commitment to an environmental ethic and encouragement for students to take responsibility for their own behaviour. This approach can be described as 'education for sustainability' (Huckle, 1993a, p. 61).

The value of recognising the distinctions between these approaches is that it draws attention to the underlying priorities and aims which teachers may have. This is crucial, because their aims are likely to be reflected in their practice and consequently in the nature of the environmental education which their students experience.

In recent years, environmental educators (for example: Huckle, 1990; Fien, 1993a) and international conference reports (for example IUCN, UNEP and WWF, 1991; UNCED,1992) have increasingly advocated education for sustainability as the most appropriate approach for meeting the world's current environmental situation. A useful definition is that given by Sterling and the Environment and Development Education Training Group (EDET) (1992, p. 2):

> . . . education for sustainability is a process which:
>
> — enables people to understand the interdependence of all life on this planet, and the repercussions that their actions and decisions may have both now

and in the future on resources, on the global community as well as their local one, and on the total environment;

— increases people's awareness of the economic, political, social, cultural, technological and environmental forces which foster or impede sustainable development;

— develops people's awareness, competence, attitudes and values, enabling them to be effectively involved in sustainable development at local, national and international level, and helping them to work towards a more equitable and sustainable future. In particular, it enables people to integrate environmental and economic decision-making;

— affirms the validity of the different approaches contributed by environmental education and development education, and the need for the further development and integration of the concepts of sustainability in these and other related cross-disciplinary educational approaches, as well as in established disciplines.

Geography teachers have the potential to embrace this process in their practice, and, in co-operating with teachers of other subjects, to make a real contribution to fundamental changes in people's attitudes and actions towards the environment. Yet a lot of support for teachers is needed if this potential is to be achieved.

Thus the original English and Welsh National Curriculum included 'few pointers to the real causes of environmental problems or to radical solutions' (Huckle, 1993b, p. 103). The new Orders are an improvement in that the concept of sustainability is specifically mentioned and environmental geography has a higher profile. At Key Stage 1 in England, the sole thematic study is 'the quality of the environment in any locality', and it includes the requirement that 'pupils should be taught how the quality of the environment can be sustained and improved' (DfE, 1995, p. 3)). At Key Stage 2, one of the four thematic studies, 'Environmental change', includes the requirement that 'pupils should be taught how and why people seek to manage and sustain their environment' (DfE, 1995, p. 6). At Key Stage 3, one of the nine thematic studies, 'Environmental issues', includes the requirement that 'pupils should be taught how considerations of sustainable development, stewardship and conservation affect environmental planning and management' (DfE, 1995, p. 14).

The National Curriculum in England and Wales, therefore, provides a clear and statutory environmental strand in geography at Key Stages 1–3, although provisions fall short of the advice given in the CCW advisory paper, *Environmental Education* (1992, p. 17) where the aims of education for sustainability are admirably summarised:

> . . . encouraging and helping people to apply knowledge and skills in wise and caring actions which reflect a growing commitment to environmental values. It is, therefore, essential that schools provide pupils with opportunities for exploring their personal feelings and responses to environmental issues, and with a climate for learning which nurtures positive attitudes towards the environment and a strong sense of social and environmental responsibility.

The National Curriculum, however, does now provide opportunities for teachers to expand and develop environmental issues in their courses provided that there is commitment and support from their schools. Furthermore, some geography teachers are already implementing schemes of work based on 'education for sustainability', some textbooks and textbook series reflect this approach, and, among GCSE and A level syllabuses, Avery Hill and Geography 16–19 respectively lead the way (see Bibliography for examples of such publications).

Teaching for sustainability through geography

How then can geography teachers implement education for sustainability, along the lines of the Sterling and the EDET group definition? In answering this question, we consider first aspects of content and then of teaching approach.

Environmental subject matter is complex and needs to be understood holistically. In the words of Johnston (1989, p. 199):

> understanding the nature of environment problems and how they might be solved requires much more than a scientific appreciation of environmental processes. It demands an understanding of how societies work and how collective action within those societies is both

organised and constrained. Only with that understanding is it possible to discuss how change might be achieved.

We believe that 'teaching for sustainability' not only involves such a holistic approach, but also necessitates the explicit recognition that these processes and interactions are viewed differently by people depending on their perspectives and value judgements. The importance of value judgements is clearly expressed by many environmental educators. Fien (1993b, p. 8) writes:

> environmental and development problems cannot be understood without reference to social, economic and political values, and . . . managing the global crisis will depend upon changes in global patterns of development and trade as well as personal values and lifestyle choices.

Posch (1993, pp. 22–23) also draws attention to the need to 'deal seriously with the search for causes' and in so doing, to have 'a vision of alternative conditions of human life'. He continues by stating that 'the absence of alternative visions is probably one of the most important obstacles to effective environmental education'.

A third area of subject matter which can contribute to 'education for sustainability' is the study of local issues. Such study, handled sensitively, allows students to generate their own knowledge, and through this, their own understanding. Students are encouraged to develop different and possibly conflicting interpretations of the causes and possible solutions of environmental issues in their locality. In this way, 'the generation of local knowledge implies an integration of experience-based judgement with available knowledge' (Posch, 1993, p. 29). This reflects the way in which teachers view the status of knowledge (Bridges, 1986), and appears to be an equivalent notion in geography to that of alternative explanations in science and constructivism in mathematics (see e.g. Driver, 1983; Jaworski, 1994).

Several content-related issues were seen by participants at the COBRIG seminar as issues of current concern to schoolteachers. These included awareness of the complexity of environmental issues especially related to students' levels of understanding; taking account of the political nature of environmental issues; integrating scales of investigation; making the most of local opportunities.

To move on from subject content to questions of teaching approach, just as we are advocating a committed approach to the teaching of environmental issues, embracing 'teaching for sustainability', we would also argue that enquiry-based learning is an appropriate approach to teaching and learning because it is most likely to provide young people with the skills they need to understand the detail of specific environmental issues and to take action as responsible citizens. Enquiry-based learning has a long tradition in geographical education; it was promoted through the former Schools Council Projects (Avery Hill, Bristol, Geography 16–19) and the Geographical Association's (GA) Geography, Schools and Industry Project and, although minimally expressed and justified, it is also advocated in the National Curriculum (paragraphs 1 and 2 in the programmes of study for each key stage in the revised Geography Orders).

Enquiry-based learning draws its rationale from models of information processing developed by educational psychologists (Kyriacou, 1986). It is characterised by a sequence of cognitive activities, frequently expressed as questions 'to provide guidelines for the ordering and selection of concepts and content within an activity or series of activities' (Slater, 1993, p. 2). It is exemplified in the 'route for learning' of the Geography 16–19 Project (Naish et al., 1987, p. 61), which consists of both factual enquiry and values enquiry. In this approach to learning, the first group of questions relates to initial observation and perception, the second to definition, description, analysis and explanation, and the third to prediction, decision-making, personal evaluation and response (Corney et al., 1992).

Enquiry-based learning is frequently advocated by geography educators, but we believe its appropriateness for 'education for sustainability' is only fulfilled in practice if it is fully implemented. It is, therefore, essential for students to answer the full sequence of questions, rather than limiting enquiry—as seems frequently to be the case—to the first two groups of questions. It is also essential that enquiry involves what are recognised in Geography 16–19 as both 'factual' and 'values' elements. If these points are taken into account, then students do have opportunities to consider the implications for the environment and society of alternative courses of action, to make judgements about the respective merits of such actions, and to decide on their own responses, behaviour and future courses of action in relation to sustainable living.

Thus, enquiry-based learning provides an effective overall approach for 'education for sustainability'. In implementing this approach, however, four points of detail should be considered. These are interrelated, and linked to the fact that environmental education by its nature is often about controversial issues.

1. A variety of strategies which actively involve students in thinking, communicating and interacting needs to be developed, carefully matched to students' existing under-standing and experiences. This progessive experience should build on foundations established in the primary school. Teaching strategies need to involve students in different forms of grouping (individual, pair, small group, whole class); in exercises promoting a range of critical thinking skills using a wide range of resources; in investigative fieldwork; in activities involving oral work, including discussion, role plays and simulated public enquiries; and in activites promoting aesthetic, intuitive, expressive and scientific experiences (see e.g. Sterling and Cooper, 1992, pp. 76–81; CCW, 1992, pp. 17–20).

2. A values education approach emphasising 'values probing' seems to be most suitable (see e.g. Fien and Slater, 1985; Slater, 1993, Chapter 4). Such an approach is more likely to be followed when the full sequence of questions in enquiry-based learning is implemented. However, we draw further attention to the area of values education not only because it needs to be an explicit part of 'education for sustainability', but also because it is an area where teachers often feel unsure about their aims and procedures.

3. The teacher's role in handling controversial issues with students needs careful consideration. The possibilities of neutrality, commitment, balance and devil's advocate are frequently discussed (e.g. Stradling et al., 1984, Chapter 1; Slater, 1993, Chapter 4). However, arguments about the nature of a balanced role in relation to teaching for sustainability are helpfully developed in Fien (1993a, pp. 87–88):

> . . . environmental education should be helping change where the balance is drawn as well as exposing the dangerous myth of value-free ideas. Instead of looking at the 'balance' involved in erroneous dichotomies such as 'economy versus environment' or 'business as usual' against 'ecocentric paths', we need to be looking

critically and imaginatively at the choices, values and problems involved in moving towards different green paths, scenarios and perspectives. . . . The balance we should be aiming for is one which restores balance within ourselves, between ourselves and others, and between people and their environments.

4.   Attention also needs to be given to the question of student response towards the end of an environmental enquiry. The actions students take in terms of their environmental lifestyle and behaviour must be their own responsibility, however much the teacher's aim is to educate for sustainable living. But teachers can take actions as part of the educational process to help their students' understanding, motivation and commitment, and to inform communities beyond the class concerned about work which has been undertaken. Some possible forms of action include:

a.   presentations to different groups of people—for example other classes, parents, students in other schools, members of the local community;
b.   displays of alternative viewpoints in school, local community centres, libraries and shopping centres;
c.   attendance at public inquiries and public meetings;
d.   letters to the press, environmental groups, politicians;
e.   direct local action: for example, improving the school environment, initiating more sustainable practices in school, cleaning a polluted section of canal bank, river or coastline.

Environmental issues in higher education

Current practice

Unlike the situation in schools, the teaching of geography at degree level is not guided by a national curriculum or any similar agreed set of national guidelines. Individual universities and departments have their 'mission statements' and geography courses are laid down by the departments themselves. There is, therefore, a wide range of geography degree courses, with some university geography departments putting greater emphasis on the teaching of environmental issues than others. In some departments, such topics constitute

core areas of the course, while in others they are offered as
options. It is also the case that certain issues, for example
human-induced global warming, will be covered as a core
topic, but studied in greater detail in an optional course on
climatology. At the same time there has also been a
proliferation, both inside and outside geography departments,
of new courses on environmental management and conservation. Many of these new courses link geography with
other disciplines, such as zoology, forestry and economics.

The content of courses

One facet of the lack of national co-ordination in the teaching
of geography in higher education is the fact that, to some
extent, the courses offered and their emphases reflect the
research interests of the staff. Certain environmental issues
(e.g. the greenhouse effect, acid rain, deforestation, soil
erosion) will be found in most environmental geography
courses. Some of these topics have a long history as teaching
subject matter (e.g. deforestation, soil erosion), others reflect
more recent global concerns (e.g. desertification since the
mid-1970s) while a few issues are more recent still (e.g.
human-induced global climate change and stratospheric
ozone depletion in the last 10 years or so).

The perspective taken when considering these issues also
varies widely. When we use the word 'issue', we imply that
human values play a key role in defining the subject matter.
However, the approaches taken to studying these issues
range from a predominantly physical geography approach
which might emphasise geomorphological processes and
rates of change, to a more humanistic view involving
perceptions, cultural contexts and political viewpoints. The
human dimension to what are, in most cases, traditionally
physical topics is increasingly being emphasised, reflecting
a discernible shift in focus at the governmental and international level. A full understanding of desertification, for
example, requires both an appreciation of the physical
processes operating in the dryland environment and a grasp
of the social, economic and political factors which affect the
everyday lives of desert inhabitants (Thomas and Middleton,
1994). In the study of soil erosion it has become clear that we
are well aware of how soil erosion works as a set of
processes, and physical scientists have developed many

methods for combating the soil erosion problem. Nevertheless, soil erosion problems still plague many world regions, suggesting that the emphasis on the technical is not sufficient. Hence, it has become necessary to ask the question: 'why is soil erosion still a problem?' The answers to this question lie in an appreciation of the way local inhabitants view the problem (some may not view it as such at all) and the factors which force or encourage people to abuse their soil resource (Blaikie, 1985).

Thus, a full appreciation of any environmental issue must consider it as a set of physical processes viewed in a social context. There is no doubting the need to understand the physical processes operating, but issues only become issues when people are involved. This shift towards a fuller understanding of what makes an issue an issue reflects the need for a more holistic approach to the subject matter. It is a shift similar to that which we are advocating in schools.

Approaches to teaching

Teaching in university geography departments is based on lectures, classes, seminars and fieldwork. Lectures are traditionally passive learning methods, the content of which is entirely dictated by the individual lecturer, in some cases aiming for a balanced overview, in others deliberately presenting a particular line of argument. Seminars traditionally involve some discussion element, based on presentations by an academic or by students guided in their approach and reading by their supervisor. Classes usually offer a more eclectic approach, in which students may be asked to prepare different sides of the argument on a particular issue, finishing with an open discussion.

These traditional methods are increasingly being supplemented by newer approaches such as role-play exercises and simulations involving decision-making. A student might be asked to assume the role of an environmental activist who has a meeting with the Minister of Transport. The student prepares arguments with the aim of persuading the Minister to scale down the national road-building programme on environmental grounds. Another example, used for investigating the various views on tropical deforestation, might involve a small class of three students, one of whom assumes the role of a Brazilian landless peasant, a second assumes the role of the Brazilian President, and a third the role of a Greenpeace international representative. Each student is asked to give a five-minute presentation on his/her perspective on tropical deforestation, followed by a round-table discussion of the issues and arguments raised.

A simulated decision-making exercise might focus on a local environmental issue, such as a proposal to develop a new gravel pit at an attractive riverside location in an area of high unemployment. Individual students might be asked to take the stance of various local interest groups (e.g. local residents, local council, Council for the Preservation of Rural England, the gravel pit operator) in a public meeting on the proposal. Their ultimate objective is to make a decision on whether or not the proposal should go ahead. In either case, the students would be asked to draw up a list of the implications for all parties involved.

In another example, the students might be asked to assume the roles of a group of international consultants, called in to suggest solutions to the problem of inadequate grazing resources in a semi-arid village grazing ground somewhere in the Sahel. The students are asked to come up with a range of possible explanations for the lack of adequate grazing, and to suggest practical ways to solve the problem.

Experience of such approaches to teaching in universities suggests that they often present students with new ways of looking at the information and knowledge they have gained from more traditional teaching methods, and they are therefore an important supplement to the established range of learning methods. These more active approaches to learning also tend to enhance the students' understanding of the practical complexities of the issues involved. Again, this development in approaches in higher education runs parallel to those which have deeper roots in schools and which seem to be particularly appropriate for 'teaching for sustainability'.

Conclusion

There is a clear difference between schools and higher education in terms of an agreed/statutory content of geography courses and the emphasis given to environmental issues. Schools have a statutory duty to teach about environmental issues, although the depth of coverage and the approach followed vary, while in higher education there is no agreed basic content and courses therefore reflect departmental interests.

The subject matter covered in the teaching of environmental issues in higher education inevitably recognises a greater level of complexity than is possible and appropriate in schools. However, the holistic nature of the issues, involving both physical geographical and human geographical approaches, is becoming increasingly emphasised both in higher education and schools, a development which we believe to be necessary and desirable. A good understanding of both physical and human processes is an essential prerequisite for the appreciation of the ways in which different people view a particular issue. The question of the extent to which schools should be providing their students with basic knowledge, which is then highlighted as an issue in higher education, or whether the issue element should be emphasised at school is one which could be usefully discussed between geographers in schools and higher education.

The trend towards teaching methods which actively involve students is more firmly established in schools than in higher education, although we feel that there is scope for developing them further at both levels. In practice, 'action in the community' approaches are more appropriate at school level than at university, simply because school students are in most cases long-term residents of the local community. Conversely, this very fact might be used to suggest that such an approach could be used to integrate more fully university students with their temporary local communities.

Geography has a long and strong tradition of integrating the human and physical aspects of the subject. In recent times, the plethora of information and new subject areas has resulted in a trend towards increasing specialisation in the research field, and to some extent in higher education teaching. It is important for schoolteachers to maintain contacts with the developments in the discipline, but it is equally important for researchers to maintain these contacts in order to be constantly reminded of the essential integrative and synthesising nature of the subject, as well as of the previous learning experiences of their students. Environmental issues are a perfect vehicle for this cross-fertilisation of ideas and approaches.

Acknowledgements

We should like to thank participants of the seminar for providing the catalyst for a number of ideas which we have developed in this contribution, and in particular, Averil Mander for reporting the seminar discussions.

Bibliography: some 'education for sustainability' publications

CCW (1992) *Environmental Education: a Framework for the Development of a Cross-Curricular Theme in Wales*. Curriculum Council for Wales, Cardiff.

Greasley, B., Ranger, G., Winter, C. and Williamson, E. (1994) *Enquiry Geography*, Books 1–3 and Teacher's resource packs. Hodder & Stoughton, Sevenoaks.

Lambert, D. (1992) *Green Pieces*, textbook and teacher's resource book. Cambridge University Press, Cambridge.

MEG (1995) *GCSE Geography Syllabus E (Avery Hill)* 1579. Midland Examining Group/Welsh Joint Education Committee.

ULEAC (1995) *Sustainability and Growth*, Unit 3.6 in the people–environment perspectives module of the Geography 16–19 A level syllabus. University of London Assessment and Examinations Council, London.

WWF (1988) *What We Consume*, teacher's handbook and curriculum units. Global Environmental Education Programme. WWF-UK/The Richmond Publishing Company, Godalming.

References

Blaikie, P. (1985) *The Political Economy of Soil Erosion in Developing Countries*. Longman, Harlow.

Bridges, D. (1986) Dealing with controversy in the school curriculum: a philosophical perspective. In J.J. Wellington (ed.) *Controversial Issues in the School Curriculum*, Chapter 2. Blackwell, Oxford.

CEE (1994) *On the Fringe of the Machine: Annual Review of Environmental Education 1993*. Council for Environmental Education, Reading.

CCW (Curriculum Council for Wales) (1992) *Environmental Education*. CCW, Cardiff.

Cloke, P., Philo, C. and Sadler, D. (1991) *Approaching Human Geography*. Paul Chapman Publishing, London.

Corney, G. J. and members of GSIP (1992) *Teaching Economic*

Understanding through Geography. Geographical Association, Sheffield.

DES (Department of Education and Science) (1991) *Geography in the National Curriculum (England).* HMSO, London.

DfE (Department for Education) (1995) *Geography in the National Curriculum (England).* HMSO, London.

Driver, R. (1983) *The Pupil as Scientist.* Open University Press, Milton Keynes.

Fien, J. (ed.) (1993a) *Environmental Education: a Pathway to Sustainability.* Deakin University Press, Geelong.

Fien, J. (1993b) *Education for Sustainable Living: a New Agenda for Teacher Education.* Background Paper for UNESCO Asia-Pacific Regional Experts' Meeting on Overcoming the Barriers to Environmental Education through Teacher Education. Grifith University, Brisbane.

Fien, J. and Slater, F. (1985) Four strategies for values education in geography. In D. Boardman (ed.) *New Directions in Geographical Education.* Falmer, Lewes, pp. 171–186.

Holt-Jensen, A. (1980) *Geography: its History and Concepts.* Harper and Row, London.

Huckle, J. (1990) Environmental education: teaching for a sustainable future. In B. Dufour (ed.) *The New Social Curriculum.* Cambridge University Press, Cambridge, pp. 150–166.

Huckle, J. (1993a) Environmental education and sustainability: a view from critical theory. In J. Fien (ed.) *Environmental Education: a Pathway to Sustainability.* Deakin University Press, Geelong, pp. 43–69.

Huckle, J. (1993b) Environmental education and the National Curriculum. *International Research in Geographical and Environmental Education,* 2(2), 101–104.

IUCN, UNEP and WWF (1991) *Caring for the Earth.* International Union for the Conservation of Nature, Gland.

Jaworski, B. (1994) *Investigating Mathematics Teaching: a Constructivist Enquiry.* Falmer, London.

Johnston, R.J. (1989) *Environmental Problems: Nature, Economy and State.* Belhaven, London.

Kyriacou, C. (1986) *Effective Teaching in Schools.* Blackwell, Oxford.

Naish, M., Rawling, E. and Hart, C. (1987) *The Contribution of a Curriculum Project to 16–19 Education.* Longman, Harlow.

NCC (National Curriculum Council) (1990) *Curriculum Guidance No. 7: Environmental Education.* NCC, York.

Pattison, W.D. (1973) The four traditions of geography. In J. Bale, N. Graves, and R. Walford (eds) *Perspectives in Geographical Education.* Oliver and Boyd, Edinburgh, pp. 2–11.

Posch, P. (1993) Research issues in environmental education. *Studies in Science Education,* **21**, 21–48.

Slater, F. (1993) *Learning through Geography,* Pathways in Geography 7. National Council for Geographic Education, Indiana.

Sterling, S. and Cooper, G. (1992) *In Touch: Environmental Education for Europe*. World Wide Fund for Nature, Godalming, UK.

Sterling, S. and the EDET Group (1992) *Good Earth-Keeping: Education, Training and Awareness for a Sustainable Future*. Environment and Development Education and Training Group. UNEP-UK, London.

Stradling, R., Noctor, M. and Baines, B. (1984) *Teaching Controversial Issues*. Edward Arnold, London.

Thomas, D.S.G. and Middleton, N.J. (1994) *Desertification: Exploding the Myth*. Wiley, Chichester.

UNCED (United Nations Conference on Environment and Development) (1992) *Earth Summit '92*, Chapter 36. UNCED, Conches.

WO (Welsh Office) (1991) *Geography in the National Curriculum (Wales)*. HMSO, Cardiff.

WO (1995) *Geography in the National Curriculum (Wales)*. HMSO, Cardiff.

The contribution of geography to personal and social education

20

David Lambert and Hugh Matthews

Introduction

At the 1994 COBRIG seminar Peter Haggett (see Chapter 2) asked us to think about the 'residuals of a geographical education: what are we left with long after the examination has been written, the results posted and geography (whether in school or in higher education) has become a fond (or otherwise) memory?' It is a good question the answers to which may surprise us. Maybe a few disconnected facts, an out-of-date map of the world, some half-digested conceptual curios, like 'K = 3' or 'stone stripes'? Or, is there something more worth while and longer lasting to be gained from what Slater (1994) calls *Learning through Geography*? Might there be, for example, an understanding of the way deeply held values guide people's decision-making, perhaps in the context of explaining land-use patterns or environmental policies?

Individuals may sometimes be able to trace back to their geography education a basis for understanding new ideas and perspectives as they emerge. Nearly a quarter of a century ago, William Bunge wrote: 'Mankind starts out trying to conquer nature then ends up learning to live with it, or else. Physical and human geography had better not be split' (Bunge, 1973, pp. 328–329). As he admitted, Bunge was not the originator of that thought, but it is an idea that seems

Geography into the Twenty-first Century, Edited by E.M. Rawling and R.A. Daugherty, © 1996 John Wiley & Sons Ltd.

in need of regular reinvention. To what extent does geography have a responsibility, if not to inculcate such ideas then at least to help establish the means for an informed level of debate of the associated issues? That is to say, to provide people with access to environmental and social matters of potentially enormous significance to their lives. What more could be meant by geography's contribution to a personal and social education (PSE)? This chapter explores in more detail some specific notions of geography's contribution to this dimension of education, first in schools and then in higher education.

Geography and personal and social education in the school

We begin by identifying *opportunities* which exist for geography in school, and more particularly in the secondary school, to contribute to PSE. This discussion does not claim to be exhaustive because to attempt to be so would require us, perhaps tendentiously, to define the precise limits to what is meant by personal and social education. Many questions arise in considering a topic which cuts across the conventional boundaries of the process of schooling. How far do free-standing courses in PSE offer an adequate range of opportunities for contributing to this aspect of a student's educational experience in school? Where does 'profiling' of student achievements and abilities fit into the discussion? How does PSE relate to vocational preparation and its emphasis on transferable skills? We adopt a broad approach to possible opportunities, identifying ways in which geography might contribute to children's intellectual growth, personal competence and skills of relating to other people, wherever such opportunities may present themselves. We then briefly consider constraints operating in schools which may impede the development of PSE through geography.

Identifying opportunities

Schools in the UK are typically concerned to fulfil educational aims which extend beyond the narrowly academic. Moreover, in spite of growing competitive pressures in the form of crude league tables of examination results (in England and

Wales), many schools are deeply committed to a concern for the 'whole child'. An interesting question raised at the COBRIG seminar was the extent to which, in secondary schools, geography departments felt part of this commitment and concern. Geography subject specialists, trained through the route of a specialised degree and the subject-oriented Postgraduate Certificate in Education (PGCE), are usually highly competent in organising the subject content for the National Curriculum and for geography courses leading to public examinations. They can be equally confident in arguing geography's corner, justifying the subject's place on the curriculum for example, or campaigning for increased timetable space and additional resources. This is done on the basis of a knowledge and understanding of geography and the claims that can be made for the subject. Thus, we hear claims made for the teaching of 'graphicacy', IT competence, environmental awareness, economic and industrial understanding, international understanding and so on; the list is potentially very long.

And yet, making such claims can seem, particularly to outsiders, somewhat unconvincing and, one senses, the longer the list is allowed to grow the greater the likelihood of it becoming mere rhetoric. Moreover, the claims for geography are dominated by the cognitive, particularly in relation to the knowledge, understanding and skills that teachers consider pupils 'need' to support preparation for public examinations. Content such as this understandably becomes the stuff of geography lessons, but it is commonly at the expense of other aspects of learning which may have at least an equal claim.

We can make three further points arising from this perspective on priorities for learning:

1. In addition to the cognitive dimension, there are emotional, behavioural and attitudinal aspects of learning. Hargreaves (1984) identified these four 'aspects of achievement', each one involving worthwhile educational attainment in its own right and, he argued, having equal status. Unfortunately this thinking has never been fully developed in schools although growing awareness of similar perspectives such as Howard Gardner's (1983, 1994) theory of 'multiple intelligences', which further erodes widely held but simplistic notions of 'ability', may yet stimulate renewed interest in such matters.

2. Schools are required to show inspectors how children's moral and spiritual development is being guided and nurtured. However, as one contributor to the discussion at the COBRIG seminar commented, 'schools are concerned but geography departments within them seem not to realise'. There is perhaps a need for the shape of the 'moral and spiritual dimension' to be delineated with greater clarity in curriculum terms, so that geographers can be guided on how such development in pupils can be supported through the subject. The Office for Standards in Education (OFSTED), responsible for school inspections in England, has published a discussion paper, *Spiritual, Moral, Social and Cultural Development* (1994), which sets out some valuable insights on this.

3. There are certain kinds of knowledge, understanding and skill that are required to underpin the sort of work being alluded to in the above; material, in other words, which forms a 'hard' context or provides a substance and content for learning. It appears, however, that much of the content of geography lessons does not necessarily 'connect' in this way to students' real lives and thus contributes to lessons becoming 'irrelevant' or 'boring' in the eyes of too many students.

An example may serve to illustrate what we mean here. In schools all over Britain, children need the appropriate time and space to examine the diversity of human populations. To do this involves more than formalising the subject matter into concepts such as 'residential segregation' or the 'push and pull factors' influencing migration. Pupils come to such topics with existing personal knowledge in the form of their own 'common sense' understandings, derived from sources outside school. These may contain a number of preconceptions which may begin to harden into prejudices, which are extremely difficult to dislodge unless (we assume) the pupil is able to accommodate and eventually assimilate new ideas into her/his existing knowledge systems. This process depends first and foremost upon the teacher openly acknowledging that pupils' prior knowledge exists; that their minds are not clean sheets awaiting imprint. It must also rely upon pupils receiving the information they need in order to make better sense of the world and for their prejudices to be challenged openly and effectively. In a sequence of lessons, then, on urban geography or migration, there exist opportunities for children to share what they know (or think they know) and, with a skilful teacher's intervention where necessary, to raise an agenda of questions. Many of these questions may be answered straightforwardly with hard information. What is the ethnic breakdown of the UK population? What are the current immigration and emigration figures for the UK? What are the origins and destinations of these groups? Beyond that, a further stage in the sequence of activities would involve encouraging students to relate their extended knowledge and understanding back to their own lives.

Geographers have an impressive track record of wide-ranging claims for the subject as a medium for education (see, for example, Bailey and Binns, 1986). Perhaps the geographical education community now needs to take an additional step, to show where the opportunities lie to serve the personal and social educational needs of students of all ages. A place to start might be to identify the knowledge, skills and understanding which students need, on matters such as health and sex education, environmental education, economic, social and political education. It should be possible for mainstream geography education, within the framework of the National Curriculum for younger children and examination syllabuses for older students, to supply the appropriate contexts for the development of such knowledge, skills and understanding (for example, a focus on economic activity might provide the starting-point for changing job opportunities for young people in the local area). Such a mission could become a practical focus for study groups across the academic communities of geographers in school and higher education.

Identifying obstacles

Thus geography can be seen as a vehicle offering opportunities for contributing to a range of cross-curricular 'themes' (such as health education) and broader, perhaps more elusive, moral and spiritual 'dimensions'. Many schools have developed free-standing courses under the guise of PSE. However, geography teachers may or may not contribute to such courses and, if they do, they have to set aside their identity of geography teacher in order to enter the distinctive 'PSE' culture. They become, as it were, non specialists and many teachers find that uncomfortable.

This general scenario has within it several sources of friction which may impede progress in any attempt to broaden education goals. First, there is potential duplication of effort. If a school has chosen a free-standing PSE route it may prove difficult in a number of ways for the geography department to undertake the sort of curriculum development necessary to turn opportunities identified into opportunities exploited. On a purely instrumental level, colleagues (including senior managers within the school) may feel that raising GCSE scores, for example, is not congruent with the development of personal and social dimensions through geography since these are matters which are 'covered' elsewhere in the curriculum. The challenge here is for us to ensure that the types of teaching emphases outlined in the previous section are not exceptional but are part of mainstream good practice. We need to believe that good examination performances follow when students feel motivated, confident and cared for; and conversely, that mediocre examination results can be produced by students who experience little more than a narrowly conceived drilling in what is perceived to be the essential subject content. It is worth noting in this context that school inspectors and others are becoming increasingly sophisticated in distinguishing the good from the mediocre or inadequate, using concepts such as 'value-added' derived from research in recent years into school effectiveness (e.g. Sammons et al., 1994; OFSTED, 1994).

A second potential obstacle, clearly associated with the above point, is the feeling sometimes voiced that interpreting geography as a medium of education with broad goals beyond the narrowly academic, carries a high political risk. The dampening effect of an over-cautious school management is all too easy to imagine and, conversely, the creativity which can be released by a confident and risk-taking management has been identified as a key component of 'school improvement' (e.g. Barth, 1990). Yet caution, being the safer option, is the strategy most commonly identified with school culture. We can speculate over possible explanations for this. The drive to greater market-place accountability may mean that headteachers give a higher priority to delivering what they perceive the market believes to be important. In the spirit of 'back to basics', this perception can be unduly coloured by concerns of national test results and league tables. This is a regrettable, though unsurprising, outcome of the state intervention and control enacted during the 1980s. As Walford (1992) has explained, the principal features of geography in the National Curriculum (in its 1991 configuration) owed much to contemporary anxieties about a fragmented and illegible curriculum, distorted by ill-defined (and ill-disciplined?) progressive curriculum movements such as integrated humanities and world studies. On the other hand, now that geography has, in a sense, 'seen off' such curriculum initiatives, geographers in schools must address the kinds of imperatives that led to these developments in the first place.

Any suggestions to take geography much beyond the statutory and nationally tested core thus runs the risk of reawakening old dragons. But we can take some heart from the Dearing Review (Dearing, 1994) and the reformed curriculum which followed, helpfully reducing the volume of content to be tested. Some see this curriculum as so slim that 'anorexic' is the only apt description, but this is unfairly pejorative and perhaps betrays an unwillingness to trust teachers to design and implement coherent and meaningful curricula. This, since the Review, they can now do without the anxiety of having to get through a closely prescribed and externally tested content. There is now an opportunity for principles of curriculum planning and design (see Rawling, 1996) to re-emerge from their forced exile since the Education Reform Act of 1988 and the subsequent specification of a detailed and overly complex National Curriculum. What this means, above all, is that teachers can identify their goals before setting out lesson-by-lesson objectives and many teachers will not be satisfied, professionally or personally, with goals which do not encompass the consideration of pupil needs and wider educational aims.

If the original National Curriculum was an obstacle (see e.g. Lambert, 1994) the revised version can in this respect be seen as an opportunity, which returns us to the earlier suggestion that geography curriculum planning and design can benefit from identification of specific opportunities, content and strategies which will include aspects of PSE.

There is a third obstacle which, while potentially impeding the development of a broader geographical education, could also possibly be turned to advantage. Even in an educational climate friendlier than the present one for teachers to employ the sorts of approaches implied in the previous section—involving students, for example, in sharing their own knowledge, or having an input in developing critical enquiry learning strategies—requires some risk pedagogically. Open-ended teaching techniques can have unpredicted outcomes. They may even 'go wrong' and in less skilful hands can yield limited educational outcomes. The same is true of traditional teaching strategies, but usually in a less visible manner and out of earshot; and, of course, it is the risk of the public failure that raises the stakes for teachers.

We are, however, at a watershed in schools (perhaps also in higher education, though for different reasons) in that there is a renewed interest in pedagogy. It is precisely because the National Curriculum provides some limits to the question of *what* should be taught that teachers can now focus much more energy on the question of *how* to teach. It is possible to speculate that, in the promised post-Dearing calm, we may see pedagogy, traditionally a blind spot in British educational studies, emerge as a key component of the art of teacher as curriculum developer. In higher education also, as the remainder of this chapter will argue, geographers are facing some interesting pedagogical challenges (see Gold et al., 1991). It seems likely that, in both sectors, teachers and lecturers will take on board aspects of PSE. Such a trend implies a more skilful teaching force, with teachers and lecturers having the capacity to develop wider and more responsive repertoires. There is here a real opportunity to cross academic boundaries to share experience and knowledge.

Geography and personal and social education in higher education

In considering PSE in higher education, discussion will focus on three issues: skills-based approaches to geography teaching; multicultural education and geography; political literacy and geography. Once again, the common thread is how geography, in this case within 'new' and 'old' universities and colleges, contributes to PSE. In each case, change of some kind is taking place, characterised in turn by proselytisation and development, redefinition and re-emergence, and repositioning and consolidation. What is clear is that geography is on the move, such that many of today's geography graduates are comprehensively equipped with a broad range of generic and subject-specific skills and a keen appreciation of the world about them. Less clear, however, is whether these skills will remain embedded within the graduate geographer, becoming what Haggett conceives as the 'residuals' of a geographical education. Questions as to the lasting effects of a geographical education would provide a fruitful field for further research.

Skills-based approaches to geography teaching

The agenda within higher education is changing. The first signs of a transformation began to appear in the late 1980s. The implementation of the Technical and Vocational Educational Initiative (TVEI) in schools was quickly followed by the Enterprise in Higher Education Programme (1988) for universities, polytechnics and colleges. Both schemes drew on the concept of 'enterprise education', which was conceived as a holistic learning experience, designed to link the worlds of education and the workplace. For the first time, the development of generic transferable skills was to be an explicit part of the curriculum. A gauntlet was thus thrown down to those who saw higher education as a liberal and emancipatory experience, sufficient in its own right as a preparation for most employment. Critics railed against the perceived threat of centralisation and corridors were full of rumour that this was the beginnings of a national curriculum for higher education, which could only be a bad thing (Clark, 1990; Bradford, 1995).

By the mid-1990s the clamour of criticism had subsided. 'Enterprise' had become part of the teaching culture of many courses with skills and content occupying complementary slots on the timetable. Perhaps part of the reason for this was that the 1990s had seen other changes which had caused academics to reflect further on the value and purpose of higher education. Three developments are of particular note. First, the market for graduate employment had changed significantly. No longer was a degree a passport to guaranteed employment and the idea of one career/job for life seemed to have become a thing of the past. There is evidence (Harvey, 1993) that employers, particularly those who are keen to encourage adaptability, co-operation, self-reliance and self-expression, place high value on transferable skills, as well as on disciplinary knowledge. Second, the nature of higher education had changed considerably with the drive towards mass participation. Average class sizes had increased enormously and with it came a redefinition of the relationship between lecturer and student.

Reflecting earlier moves in schools, there is now in higher education a growing emphasis on student-centred teaching strategies, active learning and learning as an open-ended process, and with it has come a reinterpretation of the role of the teacher from 'authority figure, through interpreter and demonstrator to adviser and colleague' (Gold et al., 1991, p. 3). Whether such innovation would have come about in any case as part of healthy pedagogic reflection, or whether it is a survival strategy of a beleaguered profession, is a matter of conjecture. Thirdly, the teaching quality assessment exercise encouraged widespread reappraisal of the delivery process and challenged the providers of higher education to reconsider issues of accountability. Both parents and students deserve quality and in the context of a difficult labour market, 'fitness for purpose' can only be defined in relation to a broad set of learning outcomes.

Geography has been well placed to respond to many of these challenges, with geographers among the most active in the process of encouraging reflection, reappraisal and realignment. Personal and social skills have become both an implicit and an explicit part of the geography curriculum in higher education.

Implicitly, geography encourages the development of a wide skill base. Unlike many subjects, the nature of geography lends itself to a variety of delivery styles to which students are customarily exposed. Students typically have experience of fieldwork, practical work, project work and laboratory work, in addition to conventional classroom teaching. The 'hands-on' ethos of the subject draws students to work and study in different environmental contexts, both within the UK and overseas. Also, given the breadth of the subject, there is a range of skills which form part of most higher education geography courses. These include numeracy, graphicacy, computer literacy, especially in relation to GIS, data collection and analysis and presentational skills of various kinds.

Explicitly, geographers have been at the forefront of developing courses which firmly place personal and social skills on the teaching and learning agenda. An important forum for the dissemination of 'good practice' has been the *Journal of Geography in Higher Education*, which was founded on the conviction that the importance of teaching had been consistently undervalued in higher education.

Other initiatives, too, have consolidated the recognition of generic transferable skills. For example, funding from the Department of Employment has supported two disciplinary networks focusing, in turn, on enterprise and transferable skills (Geography Enterprise Network—Jenkins and Ward, 1995) and student empowerment (Geography in Education Network for Empowerment (GENE), 1994–95). Behind both networks is the mission that active rather than passive learning is desirable and beneficial to both the student and the teacher. These networks promote the view that students should become both explorers of existing knowledge and creators of their own knowledge. Skills-based curricula challenge students to be critical, constructive and flexible. They place a considerable emphasis on a broad range of transferable skills and focus on the achievement of understanding and personal development.

However, these ideas have not been taken up evenly across the system and they have been criticised from two perspectives. Some see an explicit focus on generic skills as a threat to academic virtue, while others disagree with the notion of active student participation in course design and delivery. Evidence from the networks suggests that, while most of the new universities and colleges were either receptive to change or already well along the road to designing novel, student-centred curricula, many of the traditional universities were less responsive to these ideas and took a narrower view of the definition and value of transferable skills. A further point of contention is whether skills-based courses should be set-up as 'stand-alone' units, of value in their own right, or whether they should be integrated within the learning objectives of all course units, constituting a spiral curriculum of skills. As we have noted, a similar debate is taking place in schools.

A variety of student monitoring procedures has evolved alongside this growing recognition of the value of a holistic learning strategy, in which transferable skills are deemed as worthy as academic credentials. (Fenwick et al., 1992). The most widespread of these procedures are schemes which encourage students to reflect on their current level of capability, across a spectrum of skills, and to define realistic personal objectives which are testable in the course of time. In this way, students are encouraged to assess the 'added-

Table 20.1   Example of a profiling proforma

How do you rate your competence in each of the following skills?*
Please answer using the following scale:
(1) None; (2) very little; (3) reasonable; (4) good; (5) very good.

Personal/interpersonal skills

1	Learning quickly	1/2/3/4/5
2	Critical thinking	1/2/3/4/5
3	Managing time	1/2/3/4/5
4	Prioritising	1/2/3/4/5
5	Creativity	1/2/3/4/5
6	Evaluating	1/2/3/4/5
7	Developing reasoned arguments	1/2/3/4/5
8	Putting ideas into practice	1/2/3/4/5
9	Acting assertively	1/2/3/4/5
10	Appreciating own strengths	1/2/3/4/5

Communication skills

11	Communicating (oral) with colleagues	1/2/3/4/5
12	Giving presentations	1/2/3/4/5
13	Listening	1/2/3/4/5
14	Communicating by telephone	1/2/3/4/5
15	Communicating by writing	1/2/3/4/5

Group work skills

16	Working in a team	1/2/3/4/5
17	Assessing the contribution of others	1/2/3/4/5
18	Leadership	1/2/3/4/5
19	Negotiating	1/2/3/4/5
20	Resolving conflict	1/2/3/4/5
21	Delegating	1/2/3/4/5
22	Motivating others	1/2/3/4/5
23	Working independently	

Project skills

24	Identifying problems	1/2/3/4/5
25	Solving problems	1/2/3/4/5
26	Planning a project	1/2/3/4/5
27	Designing a questionnaire	1/2/3/4/5
28	Sampling	1/2/3/4/5
29	Searching for/retrieving secondary data	1/2/3/4/5
30	Selecting statistical techniques	1/2/3/4/5
31	Report writing	1/2/3/4/5

IT/numerical skills

32	Word processing	1/2/3/4/5
33	Using statistical packages	1/2/3/4/5
34	Using spreadsheets	1/2/3/4/5
35	Using databases	1/2/3/4/5
36	Using graphics packages	1/2/3/4/5
37	Drawing and cartography	1/2/3/4/5
38	Numeracy	1/2/3/4/5
39	Spatial analyses	1/2/3/4/5

*Survey undertaken at beginning and end of each academic year.
Source: Based on a profiling scheme at Coventry University (Division of Geography).

value' of their courses. Strongly linked to the notion of school-based 'records of achievement', these procedures of 'profiling' and assessing 'learning outcomes' have added to the momentum towards learner empowerment. Geographers have been quick to incorporate these procedures (Table 20.1) and there are many instances of good practice (Jenkins and Ward, 1995). These developments consolidate the view that higher education should be seen as a further building block in preparing students for a process of lifelong learning.

But there is still some way to go before the notion of autonomous learning and skills-based teaching becomes an accepted and integral part of most geography courses. For some, the challenge of 'letting go' of the learning process will always be an unsurmountable hurdle. Yet there need not be any conflict between the provision of 'content' and 'skills'. Indeed, the development of skills-based approaches can enhance and extend a student's appreciation of the intrinsic values of a subject, and lead to a fuller understanding of content and disciplinary knowledge. While students seem ready enough to take part in these challenges, there is a need to encourage an appropriate range of associated staff development opportunities. The experience of the two disciplinary networks suggests that there is a danger in always talking to the converted.

Multicultural education and geography

Not only is geography changing in its recognition of skills, but the subject is also redefining its outlooks and perspectives. This is particularly evident in its approach to understanding culture. In the recent past, culture was seen to be a unitary and élitist conception, which bound together disparate groups living within a society (Jackson, 1992). In this context, a Western culture, characterised by cosmopolitanism and progressiveness, was set against a non-Western culture, the artefacts and settings of which were the products of a lack of civilisation and technological backwardness. Stereotypical images prevailed to provide a less than satisfactory view of the world.

Today, cultural geography is in the process of redefinition and reappraisal. Its focus has shifted towards recognising that society comprises a plurality of groups, each with their own 'ways of seeing' and experiencing the world. For Stock (1983) culture is defined by groups who cohere around shared visions, languages and codes of practice. Such a definition enables many different reference groups to be recognised, whether based on socio-personal criteria (class, gender, ethnicity, age, disability, sexuality), political identity (voting behaviour, party allegiance, oppression) or environmental disposition (peace activists, conservationists). The challenge for geography is to disentangle the shared subjectivities of place in order to make sense of these alternative visions of the future (Anderson and Gale, 1992).

This is no easy task and a major problem is one of data collection. Critics of the 'old' cultural geography draw attention to its overdependence on top-down methodologies, typified by an outsider perspective. The 'new' cultural geography is built around 'construction', founded on participant observation, in-depth interviews of groups and individuals, records and minutes of meetings and community immersion of various kinds; and 'deconstruction', based on a reinterpretation of landscape and text, such that the 'silences' of place and the media begin to resonate with experiences and values.

Multicultural education is intricately bound in with conceptions of power. How society is 'presented' reflects representations of normality which often serve the interests of an élite. By the same token, popular culture and dominant ideology are social constructions which serve the interests of particular and powerful groups. Geography is increasingly sensitive to such misrepresentation and has been strongly influenced by work in the cognate disciplines of sociology, political science and cultural studies (Anderson and Gale, 1992). This coming together of thoughts and ideas has reinvigorated and revitalised cultural geography 'making it one of the most buoyant areas of human geography' (Jackson, 1992, p. 97).

Political literacy and geography

The resurgence of cultural geography has in itself encouraged a keener understanding of both the politics of place and the distribution of power within society. However, the potential for geography students to develop a keener sense of political issues has come about through the consolidation and repositioning of what was once a much neglected subdiscipline, political geography.

The political upheavals of the 1990s have reinforced a growing interest in political patterns and processes. Three levels of analyses provide the focus of the subject. Taylor (1992, p. 111) identifies these as *geopolitics*, the study of international relations and the interstate system, including geostrategic and geoeconomic conflicts; *the geography of the state*, which embraces the way in which governments and bureaucracies manage their territories; and *the political geography of localities*, which highlights the conflicts, collisions and collusions that take place within local communities. These three scales are by no means discrete and mutually exclusive and political geography attempts to highlight their interconnectedness and demonstrate the two-way nature of local and global relationships.

Political literacy, however, goes beyond the bounds of political geography. Human geography implicitly deals with controversial issues and disagreements involving values that depend upon differing explanatory frameworks. For example, whether the world is divided into 'traditional' and 'modern' societies or whether inequality is the product of a 'world core' that has underdeveloped the countries on the 'periphery' is a debate that cuts across many parts of the subject. Similarly, whether government action is regarded as 'interference' or 'protection' has much to do with urban, social and economic geography. The notion that explanation in human geography is not value-free inevitably encourages students to grapple with judgements and constructions often beyond their own experiences. Similarly, the postmodern critique poses a view of place and space which emphasises diversity. In this sense, student geographers are increasingly well prepared to unravel the complexities of social and environmental issues and to make sense of their world.

Conclusion

From our brief survey a number of common issues emerge. Firstly, 'active-learning', 'profiling' and 'records of achievement' are no longer just the 'buzz words' of the school community. In recent years there has been a growing realisation by higher education of the need to take on board and develop the concern shown by the school sector for the notions of student-centred teaching and lifelong learning.

Most schoolteachers are happy with teaching 'skills', both generic and subject-specific, through content. This has been reinforced by curriculum developments of the kind associated with the 16–19 Geography Project, which is now the dominant geography syllabus at A level. In higher education, while there is less consensus as to how the development of skills should be provided for within the curriculum, there is a growing awareness of the value of a spiral curriculum of skills. Notions of coherence and of progression are no longer confined to subject details, but embrace the way in which students are encouraged to view their learning as a holistic experience. It is interesting to note, in relation to this range of ideas about both fostering and monitoring the development of skills, the renewed focus of attention on pedagogy in schools and in higher education.

Secondly, the importance of 'values' in geographical education is shared by both the school and higher education sectors. Geography provides opportunities to open the minds of children to the complexities of their everyday worlds. Its increasing emphasis on the subjectivities of place and the personal geographies of different populations encourages children and students to see environments from the point of view of others. To this end, geography is an emancipatory subject which encourages not just an understanding of 'what, where and why', but also of 'what can I do?'.

Thirdly, another common theme emerging from both sectors is the concern to 'root' development of broader skills and abilities in specifically geographical contexts, a concern with its origins both in the need to consolidate the subject's position in the curriculum and in a genuine commitment to geography's contribution to broader educational aims. For geography to make a significant contribution in this sense, it is clearly necessary to go beyond the claims and the rhetoric and to demonstrate how 'education through geography' can make a difference to the student. Some of the possibilities referred to in higher education—in the geography of international relations, the political geography of localities, the 'new' cultural geography—could link usefully with parallel interests in exploiting geography as a vehicle for general education in schools. The potential for a fruitful dialogue is clearly there.

Finally, it has to be acknowledged that increasing central control of the curriculum could threaten the further development of geography's role in PSE. The drive towards centralised control is no longer just a concern of schoolteachers. While the demands of the National Curriculum have shackled the development of geography within schools, the demands of large class sizes and static resources within higher education is leading some (Bradford, 1995) to fear that it is only a matter time before a 'quick-fix solution' is sought, such as that provided by a standardised curriculum supported by distant learning packages. Indeed, the impetus behind the various Teaching and Learning Technology Programmes (TLTP) (Phases 1 and 2) has been to provide computer-assisted learning (CAL) freeware to be widely disseminated throughout the higher education sector. Geography has its own CAL initiative, GeographyCAL (the UK CAL consortium in Geography, TLTP Phase 2), seeking to provide a large set of teaching and learning software, spanning across human and physical geography.

Developments of this kind provide an attractive and convenient option for the harassed academic, increasingly torn between the rival demands of the research assessment exercise and teaching quality assessment on the one hand, and the consequence of the drive towards 'one in three' in higher education by the millennium on the other. Yet, the spirit of higher education has always been one of diversity of opportunity and among the strengths of geographical education in all sectors of education is the implicit recognition of the importance of variety and non-conformity.

References

Anderson, K. and Gale, F. (1992) *Inventing Places: Studies in Cultural Geography*. Longman Cheshire, Melbourne.

Bailey, P. and Binns, T. (eds) (1986) *A Case for Geography*. Geographical Association, Sheffield.

Barth, R. (1990) *Improving Schools from Within*. Jossey-Bass, San Francisco.

Bradford, M. (1995) Common core or food for the political worm. *Journal of Geography in Higher Education*, **19**(1), 5–9.

Bunge, W. (1973) Ethics and logic in geography. In R. Chorley (ed.) *Directions in Geography*. Methuen, London, pp. 317–331.

Clark, G. (1990) Enterprise in higher education. *Geography*, **75**(2), 141–144.

Dearing, R. (1994) *The National Curriculum and its Assessment: Final Report*. School Curriculum and Assessment Authority, London.

Fenwick, A., Assister, A. and Nixon, N. (1992) *Profiling in Higher Education: Guidelines for the Development and Use of Profiling Schemes*. HMSO, London.

Gardner, H. (1983) *Frames of Mind*. Basic Books, New York.

Gardner, H. (1994) The theory of multiple intelligences. In B. Moon, and A. Shelton Mayes (eds) *Teaching and Learning in the Secondary School*. Routledge, London, pp. 38–46.

Gold, J., Jenkins, A., Lee, R., Monk, J., Riley, J., Shepherd, I. and Unwin, D. (1991) *Teaching Geography in Higher Education*. Blackwell, Oxford.

Hargreaves, D. (1984) *Improving Secondary Schools*. Inner London Education Authority, London.

Harvey, L. (1993) Employer satisfaction: interim report. Paper presented to Quality in Higher Education Conference, University of Warwick, 16–17 December.

Jackson, P. (1992) Social and cultural geography. In A. Rogers, H. Viles and A. Goudie (eds) *The Student's Companion to Geography*. Blackwell, Oxford, pp. 97–103.

Jenkins, A. and Ward, A. (1995) *Developing Skill-based Curricula through the Disciplines: Case Studies of Good Practice in Geography*. SEDA, London.

Lambert, D. (1994) The National Curriculum: what shall we do with it? *Geography*, **79**(1), 65–76.

OFSTED (1994) *Spiritual, Moral, Social and Cultural Development*. HMSO, London.

Rawling, E. (1996) The impact of the National Curriculum on school based curriculum development in secondary geography. In A. Kent, D. Lambert, M. Naish and F. Slater (eds) *Geography in Education: Viewpoints on Teaching and Learning*. Cambridge University Press, Cambridge, pp. 100–132.

Sammons, P. , Thomas, S., Mortimore, P. Owen, C. and Pennell, H. (1994) *Assessing School Effectiveness: Developing Measures to put School Performance in Context*. Office for Standards in Education, London.

Slater F. (1994) Education through geography: knowledge, understanding values and culture. *Geography*, **79**(2), 147–163.

Stock, B. (1983) *The Implications of Literacy*. Princeton University Press, Princeton.

Taylor, P. (1992) Political geography. In A. Rogers, H. Viles and A. Goudie (eds) *The Student's Companion to Geography*. Blackwell, Oxford, pp. 111–116.

Walford, R. (1992) Creating a National Curriculum: a view from the inside. In D. Hill (ed.) *International Perspectives in Geographic Education*. Center for Geographic Education, University of Colorado, Boulder, pp. 89–100.

Conclusion

New perspectives for geography: an agenda for action

21

Richard Daugherty and Eleanor Rawling

This book is a direct outcome of the 1994 COBRIG seminar. Its purpose is to continue the dialogue between schools and higher education begun on that July weekend. Individual contributors have summarised developments and trends in aspects of geography and geographical education. They have also identified issues for continuing debate and for possible future action. Co-operation between school and higher education geographers has resulted in the identification of some common concerns and suggestions for further collaboration. This concluding chapter firstly draws together the threads of a wide-ranging debate, then highlights the key issues in the next two sections, suggests an agenda for the future in the penultimate section and concludes by drawing attention to the changing context for geographical education in the early years of the twenty-first century.

Drawing the threads together

The educational contexts within which higher education geographers and school geographers work are in many ways very different. Different again, and hardly featuring in this book, is the further education sector with its more direct links to business, industry and the world of employment. As at the seminar, discussion in this book focuses on schools, in

Geography into the Twenty-first Century, Edited by E.M. Rawling and R.A. Daugherty,
© 1996 John Wiley & Sons Ltd.

particular secondary schools, and higher education. Further education and primary education are, however, being drawn more fully into the debate at the second COBRIG seminar in 1996.

In the higher education sector a range of institutions, diverse in size, in traditions and in courses, now uses the title 'university'. But all share, to a greater or lesser extent, the experience of a mix of research and teaching activities, of growing numbers of students to cater for and of an increasing preoccupation with satisfying external requirements, whether they be in relation to the quality of teaching or the quality of research. Geographers in higher education are primarily concerned with the subject, with advancing the frontiers of research and with teaching their own specialist aspect of geography. As the contributions to Part B of this book make clear, there are many research developments in particular areas of geography, but few people trying to develop an overarching picture of the subject. There would also appear to be fewer opportunities for specialists to talk to each other than there were in the mid-1970s. The pressures of research assessment ensure that the minds of most university geographers are focused on depth to ensure a strong position in research ratings and the associated funding decisions.

In schools, there is also an increasingly diverse range of institutions catering for secondary age pupils, with grant-maintained schools and technology colleges now cutting across the more obvious state/private divide. Conversely, the establishment of the National Curriculum, of the GCSE Criteria and of the A/AS Subject Core for Geography, have resulted in less diversity in what is offered as schemes of work and syllabuses are designed to fit the more prescriptive, centrally controlled curriculum guidelines. Research trends in the subject are not a pressing concern for geography teachers in schools; they are no more than a backcloth to the topics and themes listed in curriculum guidelines and syllabuses. In the late 1990s, the hard-pressed teacher is more likely to be concerned with managing statutory assessment requirements, contributing to the school's performance in league tables and meeting the demands of school inspections than with following up new ideas in geography.

The situation across schools and higher education is, therefore, that geography is taught within a great diversity of institutions and contexts. This may have the effect of impeding the lines of communication and contact between the two sectors as well as making for a more substantial divide for students to cross. Until recently this diversity of situation has not prompted any significant broadening of the debate about the educational purpose and character of the subject. Instead there has been a distinct narrowness of vision in both sectors, as institutions and individuals have concentrated on justifying funding, implementing central directives and maintaining positions in league tables and research ratings.

Identifying the divide

It is no surprise that contributors to this book identify a substantial and growing 'divide' between the two sectors. Binns (p. 37) refers to the lack of involvement by higher education geographers in the activities of the Geographical Association (GA) and quotes Goudie (1993, p. 338) in relation to 'the chasm [that] has developed between those who teach at school and those who teach in universities'. Bradford (p. 280) analyses the changing nature of the school–higher education interface, referring to course content and teaching approaches as well as to the increasing diversity of institutions and students on either side of it, while Unwin (p. 23) bemoans the resultant 'demise of university influence on primary and secondary curriculum design'. It is important to appreciate that this growing divide, although exacerbated by the current situation, did not originate in it, but may be traced back to the late 1960s and early 1970s.

The 1960s saw the heyday of positivism in university geography, with the emphasis on model building, theory and quantification. Great enthusiasm for the new ideas and approaches which these made possible was communicated to schools by means of conferences, courses and eventually textbooks. The relationship between higher education and schools was close, if paternalistic, with academics gladly sharing the 'new geography' with their junior partners in schools. It was also a relatively simple functional relationship, since schools provided the students for geography degree courses and for the next generation of geography teachers, and higher education geographers were closely involved in designing the A level syllabuses, setting the A level examination papers and writing the appropriate textbooks.

But changes were already under way in the late 1960s as increasing dissatisfaction with the narrowness of a subject dominated by a positivist philosophy led to a flowering of alternative approaches in the universities such as humanistic geography, welfare geography and radical geography. These new ideas were hardly considered by schools, partly because the process of making changes to examination syllabuses and examinations is a lengthy one and the new breed of positivist syllabuses had only just been accepted into the system in the early 1970s, and partly because schools as a whole were coming under a different set of pressures. The comprehensive reorganisation of schools which was gaining momentum after 1965 gave many secondary schools a much wider range of pupils to cater for. It accentuated the feeling that it was time to stand back from an academic view of subjects, certainly for 11–16-year-olds, and to consider the character and purpose of the school curriculum more broadly. Eventual entry into higher education was not the main factor to consider when planning curriculum content.

A range of new ideas about curriculum, many of them deriving from the USA originally, were being discussed, rethought and reapplied in the UK (Lawton, 1973). The establishment of the Schools Council in 1964 meant that there was a specific body in England and Wales charged with the task of overseeing and promoting curriculum reform and development. Like many other school subjects , geography moved into a period of curriculum innovation. The three major geography curriculum projects of the 1970s and 1980s, all sponsored by the Schools Council, ensured that the focus of attention in schools was more on educational matters, on aims, objectives and pedagogy and on the potential of geography as an educational medium. Some attention was given to the changing nature of the subject at the research frontier. For example, there are certainly signs of the influence of behavioural geography, welfare geography and even radical geography in the themes chosen by Geography for the Young School Leaver and Geography 16–19. However, there was an increasing feeling among geographers in schools that, as the 16–19 Project made clear, 'since geography is to be used as a medium for education, there is no requirement that all new academic developments necessarily be translated into the school context' (Naish et al., 1987, pp. 26–27). In any case, as Jackson shows (p. 79), the

1980s was a period of increasing diversity of focus and approach in academic geography, with research specialisms reaching out to embrace neighbouring disciplines, rather than developing internal linkages.

By the mid-1980s the divide between geography in schools and in higher education was more marked than it had been two decades earlier. The wish expressed by one of the contributors to the GA's 'Geography into the 1980s' conference that the work of the projects could provide an agenda for wider debate (Naish, 1980, pp. 61–66) was unfulfilled as geography teachers in all sectors of education were caught up in the many changes and developments of that decade. Effectively, the National Curriculum, national testing arrangements, the research assessment exercise and changes in funding at all levels ensured that the debate would not be rejoined, at least in the short term.

The need for dialogue

Given this situation and the obvious differences between the two sectors in their priorities and concerns, why is it important to engage in dialogue and what should be the matters for discussion?

A functional relationship

Firstly and most obviously, there remains a functional relationship between higher education and schools, with each supplying the other's 'seedcorn', as Walford describes it (p. 141). Schools need geography teachers who are well versed in the subject and higher education relies on schools to maintain the supply of geography students. In a practical sense, then, the two sectors are mutually dependent and it is in the interests of both to make the transitions, of students into higher education and of teachers into schools, as smooth as possible.

Geographers in higher education need to know about, and to influence, changes in the school geography curriculum in order to achieve an acceptable progression into higher education. The situation has been particularly acute in relation to physical geography, as Davidson and Mottershead explain (pp. 317–318), with the lack of a grounding in schools having led to a situation in which some undergraduates are unable to cope with degree course content in physical geography. Similarly, school geography teachers need to be familiar with recent advances in geographical understanding and knowledge if they are to provide a sound and motivating base for further work. Healey and Roberts suggest (p. 302) that recent developments in political and economic geography might be particularly relevant to school geography, and Jackson's chapter (p. 85) draws attention to the importance of making connections across scales and over time. He refers to recent work by Massey, Harvey and Smith which has direct relevance to interpreting National Curriculum requirements for 5–14 geography.

But, in this functional relationship between higher education and schools, it is not just the geographical content of courses which needs to be discussed. Smoothing the path of students who wish to take further their study of the subject requires attention to the changing circumstances in which they are learning geography. Bradford argues (Chapter 16) that a smooth crossing of the school–higher education interface also depends on mutual understanding of matters such as class sizes, teaching methods and styles, resources used, teacher–pupil relationships, course opportunities and career paths.

Considerable advances have been made recently in both sectors in some of these respects. For example, Burkill (p. 224) explains how a minimum entitlement of information technology (IT) skills and competencies has now been established for school pupils aged 5–16 and is being supported by resources and advice from the Geography/IT Support Project (GA/NCET). Although take-up of the entitlement idea is far from complete in schools, Burkill suggests that its very existence necessitates further dialogue with geographers in higher education where a different approach to IT development is under way in the Teaching and Learning Technology Project.

Newly developed vocational courses are also making an impact on both higher education and schools. In many cases NVQ and GNVQ units, being taken up by students from ages 14–21+, have some element of geography content within them (for instance, leisure and tourism, business studies). Parallel developments are taking place in Scotland (see Hunter, p. 242) with some SCOTVEC modules incorporating elements of geography into programmes with a vocational focus. It is, as yet, unclear whether these courses will compete directly with geography or provide extra opportunities for geographical work to feature (see Butt, p. 183). In order to influence the outcome, teachers in schools, in further education and in higher education need to clarify the opportunities for geography to make a distinctive contribution to vocational education, and to identify the potential pathways for students with these qualifications. The interim outcomes of the 1995 Dearing Review of 16–19 Qualifications (Dearing, 1995) reinforce the need for geographers at all levels to have thought through these matters.

Coherence in purpose and content

While the first reason for dialogue derives very directly from the needs of the students who transfer from one sector to the other, the second reason is much more to do with the subject as it is understood and interpreted in each sector. If the subject has an internal coherence and a clear identity and purpose within education, then this surely needs to be apparent at whatever level it is being taught. Haggett refers (p. 17) to conserving and re-emphasising 'some central and cherished aspects of geographical education: a love of landscape and of field exploration, a fascination with place, a wish to solve spatial conundrums posed by spatial configurations'. To this might be added the potential of geography to promote a deep understanding of the state of the environment and of our own personal responsibility for its maintenance as identified by Lambert and Matthews (p. 340) and Corney and Middleton (p. 327). These are some of the fundamental aspects of geography which should underlie all geographical education from kindergarten to degree level and beyond into adulthood, precisely because they constitute geography's contribution to understanding the world around us. They should be mutually agreed, transparent and clearly recognisable by the general public as well as by those in education.

Failure in the past to agree about fundamentals can be blamed for the misunderstandings about place and locational knowledge in the National Curriculum, resulting in locational knowledge being written into the original 1991 Order in an overprescriptive way. Johnston (p. 72), Bradford (p. 284) and Rawling (p. 260) suggest that the role and character of place

study is a central aspect of geography deserving greater
attention if the subject is to retain its vigour and identity.
O'Riordan (p. 114) is concerned that geography in higher
education has already failed to establish clearly its environmental credentials in the eyes of the general public, fund-
holders and government. Nevertheless it is still seen as an
important vehicle for environmental understanding at school
level (Corney and Middleton, p. 324, and Rawling, p. 259).

The experience of teaching and learning geography

A third reason why schools and higher education should
engage in dialogue concerns their shared commitment to
educating young people. At the most obvious level, there are
experiences to share about designing curricula and planning
detailed schemes of work and teaching/learning activities.
Higher education geographers no longer deal with a relatively
homogeneous group of able students but cater for a widening
range of abilities in larger classes. In this respect, as Haggett
suggests (p. 13), 'academicians have much to learn from
schools' both as a result of the project-based curriculum
activity during the 1970s and 1980s, and because of recent
experience in implementing National Curriculum guidelines.
Schoolteachers have become adept at 'filling out' bare
curriculum guidelines and forming them into balanced
courses, at translating statements of aims into key objectives
and teaching ideas, and at designing classroom strategies
suitable for mixed-ability groups.

School geography classrooms, as the contributors to Part
D point out, can be exciting learning environments where
pupils engage in active enquiry, participating in activities
like role play, simulation and decision-making as well as
more conventional teacher-directed classroom work. Geographers are also to the fore in higher education in varying
the conventional, limited diet of university teaching methods,
a fact borne out by the early evidence from the teaching
quality assessments. However, now that geographical enquiry
approaches are required throughout the 5–19 curriculum
and beyond, it would be helpful if educators at all levels
could reach agreement about the term and identify ways in
which progression can be built into enquiry skills and
approaches (Rawling, p. 262). Daugherty (Chapter 12) argues
that the question of progression in learning is one which we

have scarcely addressed either in schools or in higher education. He draws attention to the various attempts that have been made to identify and measure progression in geography, and establishes a tentative agenda for further research. Application of the results of such research will be needed if young people are to move easily from one sector of education to the next, and to take with them clearly identifiable gains as a result of an education in and through geography.

Transferable skills and education through geography

A fourth reason for dialogue is the increasing emphasis in both sectors on subject learning as a means to educational goals beyond the realm of subject-specific knowledge and skills. As several of the contributors to this book point out, higher education is no longer focused only on subject-specific study and training, but increasingly addresses the need to develop a wide range of transferable skills and abilities and to promote the social and personal growth of young people. Paradoxically, while in the schools sector the National Curriculum has retreated to a more subject-focused minimum entitlement, leaving individual schools to decide how to create the broader educational picture (Rawling, p. 255), a combination of enterprise initiatives, changes in students and the demands of the teaching quality assessments has resulted in higher education geographers giving more attention to these general educational objectives (Lambert and Matthews, p. 347). The agenda for dialogue in this area is becoming clearer, but so are the dangers of overstating the claims. The contribution of geography must be clearly rooted in the subject, building on its distinctive areas of knowledge, understanding and skill development.

Many of the contributors to this book have identified some contexts which modern geography offers for giving young people 'access to environmental and social matters of potentially enormous significance' (Lambert and Matthews, p. 340). The political geography of localities, the focus on insider perspectives on place, and the appreciation of global–local linkages all present specifically geographical starting-points for developing political literacy, cultural sensitivity, economic understanding and global awareness.

Geographers find out about the world using a distinctive mix of traditional mapping, fieldwork and observational skills combined with a critical questioning approach. The school curriculum projects built on this base and applied insights from educational psychology to provide enquiry teaching and learning approaches that are integrated with young people's experience, directly relevant to the real world and potentially leading to personal commitment and action. As Lambert and Matthews suggest (p. 349), with this kind of approach young people of all ages can become 'explorers of existing knowledge' and 'creators of their own knowledge'. This emancipatory view of the subject's potential, linked to the everyday contexts of locality, place, region and country which geography deals with, provides one of its strongest justifications as a medium for education in the broadest sense. Whether the debate is about 'educating for sustainability', 'personal and social education' or 'preparation for employment', geographers should be articulate in making this case.

Significantly, the potential of geographic education defined in this way may also help to justify geography's presence in the changing 14–19 curriculum. The success of the new vocational courses, the new qualifications structure proposed by Dearing, and the extent to which academic and vocational are accorded 'parity of esteem' all remain as unknown quantities in the mid-1990s. However, it is being argued by some that the best way to emphasise the continuum of learning and to reduce the divide between academic and vocational is to focus on the characteristics of high-quality learning (RSA 14–19 Report, Crombie-White et al., 1995). The RSA Report refers to the need to 'put traditional subjects at the service of the personal and vocational interests of the learner' and not to 'set them apart' (p. 61).

There is thus no shortage of topics for geographers to discuss and the preceding chapters have highlighted these in many different ways. Behind the all too obvious divide between the sectors it remains as true as it ever was that geographers draw their ideas from common origins, ideas which are constantly being renewed at source as the research frontier moves on. Whether a geographer's main task is to work at that frontier or whether he or she is drawing on such insights to open up the horizons of a pupil in school, those engaged in teaching at every level in the system share a commitment to geography as a vehicle for education. The educational missions of universities and of schools will retain their distinctiveness. There are also clear differences in emphasis within each sector in the extent to which the subject is structured and studied in its own terms as opposed to it being drawn upon to contribute to broader educational aims. But the recognition of such distinct missions and differences in emphasis should not be allowed to obscure the common issues for geographical education which can be found running through what have, until now, been largely separate debates.

Recreating the conditions for dialogue: an agenda for action

The obvious question which then arises is how can the dialogue be re-created. In response it is argued here that we need to consider the nature of the schools/higher education relationship, to look at how mutual understanding can develop further and to suggest arrangements which can be put in place, both within and between sectors, to facilitate dialogue. These points are dealt with in turn, providing an agenda for action by geographers.

Redefining the relationship

It is no longer appropriate to think of schools and higher education as the junior and senior partners in teaching geography. The mid-1990s are very different from the mid-1960s in this respect, and the new relationship is best seen as a partnership of professionals, each with different skills and expertise to bring to the debates and each with the common purposes of promoting the subject and ensuring a high quality of education through geography. The explicit recognition of this relationship will influence structures like committee membership, foci for subject association work and representation on national and international bodies. In the meantime a less definable but equally significant change in attitude should permeate and facilitate all levels of discourse about geographical education.

There is ample, if patchy, experience to build on here. For example, some university geography departments already work in a collaborative way with staff in local schools and take liaison with schools seriously. In initial teacher education there is now widespread acceptance of a relationship of equals between geographers in schools and in teacher education departments as partnership schemes of initial training have become established in the 1990s. Perhaps some of the more problematic 'frontier zones' in the world of geographical education are those within each of the sectors. To what extent are those responsible for the subject in primary and secondary schools treating each other as equal partners in the task of educating all children from 5 to 14? How often and how effectively do geographers in university education departments communicate with their counterparts in geography departments in the same institutions?

Developing understanding of similarities and common purposes

A necessary preliminary to an increased level of debate and joint action is the need to understand each other better. The COBRIG seminar and this book are contributions to this process, but it needs to be continued through all appropriate channels, nationally and locally. It is not just a question of schoolteachers knowing more about research findings and about courses and structures in higher education, or of higher education geographers updating themselves about changes to the school curriculum, although all of these are useful. As the contributors to this book make clear, it is a deeper dialogue which is now required, about the character of the subject and its potential to educate the citizens of the twenty-first century. Areas for exploration and discussion suggested in these pages include the following:

1. Which ideas from the research frontier are relevant and appropriate for school geography?
2. What distinctive knowledge, understanding and skills can geography contribute to vocational courses?
3. What are, or should be, the 'residuals' of a geographical education?
4. How can we better develop our students' ability to adopt appropriate modes of enquiry in their study of the subject?
5. How can we define and measure progression in geographical understanding?
6. How can we ensure that geography's contribution is recognised and understood by the general public and educational decision-makers?

Such a deeper dialogue calls for more extended opportunities for geographers across the sectors to give such matters the attention they require. The existing contexts for full debate of complex cross-sectoral concerns are few in number. Where these matters are aired at all, it tends to be within one of the geographical societies with a focus either on schools or on higher education. The 1994 COBRIG seminar was an all too rare occasion on which they were considered in some depth, if only briefly, across the sectoral divide.

Creating the conditions for reflection and rethinking within each sector

Before we can hold meaningful debates, we must ensure that appropriate conditions exist for creative reflection within each sector of education. It is easier to identify the constraints than to overcome them and this chapter began by highlighting some of the difficulties in the circumstances in which school and higher education geographers now work.

In schools, the intended increase in professional freedom provided by the 'slimming down' of the National Curriculum could become more apparent than real given the many constraints on time and budgets and the all-pervasive influence of national assessment and public examinations. Opportunities for professional activity have to be created, for example through the GA, with its new emphasis on regional activity, or through enterprising co-operative arrangements between LEAs as has been occurring in north-west England. Other possibilities for providing 'creative teacher time' may depend on new initiatives between the GA or AGTW and RGS/IBG and on the lobbying of government, locally and nationally, to ensure that some in-service funding allocations (the GEST budget) are targeted on geography. For the moment, it looks as if the grants for primary geography courses will continue into 1995/96, but the general trend is for central funds to be less closely targeted, and geography could suffer if this continues. Already, the developments which have taken place in geography and IT may be threatened because of changes to the government's in-service grant arrangements from 1995. Creative time and space must be found within the school sector, if further innovation and development of geography in schools are to occur.

In higher education, staff are also coping with new pressures and changing structures, although some of these, such as increasing student numbers and the demands of teaching quality assessment, can be seen to be encouraging a greater breadth of debate. The need here is for the importance of such broader debates to be recognised and valued equally, alongside the more subject-focused imperatives. Currently, geographers who concentrate on innovative teaching and on writing about these strategies, or who publish school texts, are not awarded the same status as their 'purer' subject research-oriented colleagues, and this fact will inhibit further joint activity with schools unless it is addressed. The teaching/research dilemma is not only apparent at the level of individual geographers; Johnston (1995) draws attention to a possible future scenario, driven by the HEFC funding mechanisms, in which research and teaching are separated and take place in different institutions. Most geographers are agreed that this would be detrimental to the subject; it would certainly reduce the likelihood of individuals participating in constructive debate and co-operation with the school and further education sectors.

In both sectors there is some scope for individuals and for geography departments to order their own priorities so that a dialogue about education is seen as an integral part of the work of staff rather than as an optional extra. But much will also depend on how far those responsible for funding and monitoring the system recognise the importance of time and of opportunities for regular and effective communication between colleagues, particularly those working on related matters in different institutions and in different sectors. Is anyone in government recognising that lively, relevant geography teaching in schools is not something which will happen automatically, after the initial burst of activity in preparing for and implementing the National Curriculum? On another point, where in the system is encouragement being given to schoolteachers to update and rethink their understanding of geography so that they can ensure their teaching is appropriate for the next century? And, for geographers in higher education, are those who make the 'rules' governing recognition of publications aware of, and content with, the way in which the current situation discourages writing which is directly accessible to teachers and pupils in schools? Given a broader definition of what is considered worthy of recognition, many university geographers could make a more significant contribution to the publications used in schools.

Providing channels for professional dialogue between sectors

In the 1960s and 1970s, the channels for school–university exchange were clearly defined by the A level geography examination system and university geographers made a substantial impact (see Hall, p. 146). In the new context of the 1990s, lines of communication have to be re-created. At one level, it is still important to ensure that higher education geographers sit on syllabus approval panels and mark GCSE and A/AS examination papers, and this will require higher education geographers to be available for such activities.

The broader debates require different channels of communication. At a national level, the Council of British Geography (COBRIG) is well placed to encourage liaison and information exchange among its member organisations and to set agendas for further action. As a direct result of the success of the 1994 seminar, it is now planned to hold such an event at two-yearly intervals with the 1996 seminar focusing on the quality of teaching and learning in geography.

More specific activities may best be addressed by co-operation between two or more subject/professional associations and, for this reason, it would seem desirable that there is cross-membership of committees and working groups. To a certain extent this is already occurring (often as a result of individual contacts), for example between the Higher Education Study Group of RGS/IBG and the education committees of this body and the GA. However, there is scope for it to be put on a more formal footing and to include committees focused on specific aspects of work (e.g. IT, fieldwork). The emerging regional structures of the GA, AGTW and RGS/IBG also provide a vehicle for co-operative events and joint activities, and there is potential for specific 'big initiatives' (e.g. Geography Awareness Week, 1996 organised by the GA) to be used to encourage local schools–higher education contact.

Several contributors to this book have referred to the importance of making research findings from one sector available in easily accessible form to those practising in the other sector. Healey and Roberts (p. 303) refer to 'mediators', people who have the ability to translate, interpret and communicate new geographical ideas to colleagues in schools, but this concept is equally applicable to ideas from educational research and from schools which may be relevant to geographers in higher education. A key condition for the success of the 'mediator' idea is that this mediation work is valued in professional and career terms. Healey (1994) has detected some shift in perceptions in higher education as a result of teaching quality assessments and this may become more pronounced.

For schoolteachers, strategies could include a broadening of properly funded professional development opportunities, including MA/M.Ed. courses, and the promotion of school/higher education publication ventures. The National Council for Geographic Education in the USA has recently pioneered a new form of continuing professional development certificate for geography teachers (NCGE, undated paper) which recognises and accredits a wide range of activities. An article in the newsletter of the Association of American Geographers (Bampton and French, 1995) also suggests that university geography departments in the USA should design geography 'updating modules' specifically for schoolteachers and count completion of these towards a higher degree. Similar ideas could profitably be explored in Britain.

Conclusion: a changing context for a changing geography

An agenda for action has been identified in the previous section, but it is important to emphasise that such actions will be undertaken in the changing context of the late 1990s and early twenty-first century. It is never easy to anticipate change and to prepare for an appropriate response. It is possible, however, to recognise trends in the education system which geographers can attempt to interpret and to respond to in a creative, proactive way.

Neither in schools nor in higher education is the curriculum now the private preserve of the professional, the 'secret garden' to which Minister of Education David Eccles drew attention as long ago as 1960. While the 1988 Education Reform Act marked a clear turning-point for schools in this respect, it is equally obvious in higher education institutions

(HEIs) that the geographer is no longer entirely free to devise and deliver a curriculum of his/her choice. Professional educators have had to learn to explain and to justify that which they believe to be appropriate in curriculum design. Geographers at all levels will continue to be expected to enter into dialogue with all those who have a direct interest in the outcomes of a geographical education—parents, school governors, the students themselves.

That process can sometimes be concurrent with the design and delivery of a curriculum, but increasingly it is also becoming both a preliminary to the work of the teacher and also an exercise which the geography teacher can expect to engage in after the curriculum has been developed and implemented. In the former respect, the experience of the Geography National Curriculum Working Group was a defining point in the trend towards non-geographers exerting influence in curriculum matters. Finding an appropriate discourse, to satisfy both the professional immersed in the subject and the interested lay person appointed to represent a wider public interest in the curriculum, was a challenge which at times clearly proved problematic for the Working Group. Whether the appointment of such a 'mixed' group was the right way to try to establish common ground between professional and public remains open to question. Indeed, it is tempting for geographers, still smarting from the crudely applied 'red pencil' approach of the then Secretary of State of Education Kenneth Clarke to the Geography National Curriculum in 1991, to hark back to the days when politicians and the public left things to the professionals. But it is unlikely, and perhaps also undesirable, in a climate of increased openness and accountability of public services, that the next review of the National Curriculum, scheduled for the year 2000, will revert to a curriculum design process which leaves all the main decisions to geography educators.

Over the same period the exposure of the geography curriculum in higher education to outside scrutiny is also likely to continue. The first cycle of teaching quality assessments required departments of geography, as never before, to present a curriculum rationale to peers from other institutions and to demonstrate that the curriculum as experienced by students was appropriately constructed to achieve specified aims and objectives. As yet, there is no formal, officially backed procedure to involve non-geographers in curriculum design in higher education, and to many such a procedure would represent an unacceptable challenge to fundamental principles of a university education. But no university geography department can afford to ignore the messages, originating from employers of graduates but strongly supported by government, about the need to ensure that the degree course curriculum is designed with the employability of students in mind (Authers, 1995).

The 1990s' emphasis on external accountability is also very much in the minds of schools as they try to work within the 'framework for inspection' prepared by OFSTED in England and OHMCI (Wales) to govern the regular inspection of primary and secondary schools. Here, as in university departments' responses to teaching quality assessments, there are dangers in geographers becoming too preoccupied with the ground rules laid down by those whose job is monitoring provision. Instead, the requirements of external accountability should be taken in their stride by geographers, in both schools and universities, who must retain their own clear sense of educational purpose and definition of what constitutes a quality geographical education.

For schools the experience of the National Curriculum has shown that high-quality professional debates and effective innovation will only take place where a balance between central direction, external accountability and professional initiative is clearly understood and sensitively maintained. Schools have undergone a period of considerable upheaval and trauma before reaching the post-Dearing situation in which the professional role of teachers appears to have been more fully recognised. Moves to increasing central control in higher education will need to be addressed with this experience in mind if the potential for further rethinking of the curriculum and for exchange of ideas among geographers is not to be crushed before it has even been fully developed.

The contributions to this book have demonstrated that geography is a subject with considerable potential as a medium for educating citizens of the twenty-first century. Ultimately, however, progress in all this will depend on our joint efforts to recognise this potential, to create the conditions conducive to dialogue about it, and to guard the professional freedom necessary to put it all into practice in the changing educational context of the next century. To return to David Harvey's nautical metaphor (quoted by Haggett p. 16), steering a course into the next century will not be easy, but it is hoped that the contents of this book at least provide the navigational charts required for the journey.

References

Authers, D. (1995) A degree of quality assurance. *Financial Times*, 20 February, p. 7.

Bampton, M. and French, R. (1995) Improving geographical education: a modest proposal. *Newsletter of the Association of American Geographers*, **30**(1), 5.

Crombie-White, R., Pring, R. and Brockington, D. (1995) *14 to 19 Education and Training: Implementing a Unified System of Learning.* Royal Society of Arts, London.

Dearing, R. (1995) *Review of the 16–19 Qualifications Framework: the Issues for Consideration.* Central Office of Information, London.

Goudie, A. (1993) Schools and universities: the great divide. *Geography*, **78**(4), 338–339.

Healey, M. (1994) Geography and quality assessment. *IBG Newsletter*, 2–4.

Johnston, R. (1995) Nasty, brutish and short. *The Times Higher Educational Supplement*, 9 June, pp. 14–15.

Lawton, D. (1973) *Social Change, Education Theory and Curriculum Planning.* University of London Press, London.

Naish, M. (1980) Geography into the 1980s. In E. Rawling (ed.) *Geography into the 1980s*. Geographical Association, Sheffield, pp. 61–66.

Naish, M., Rawling, E. and Hart, C. (1987) *Geography 16–19: the Contribution of a Curriculum Project to 16–19 Education.* Longman for SCDC, London.

NGCE (National Council for Geographic Education) (undated paper) *Advanced Professional Certification in Geography.* NCGE, Indiana, Pa.

Appendices

Appendix 1:
The Council of British Geography

The Council of British Geography (COBRIG), established in 1988, is the organisation which represents those societies and bodies concerned with geography and geographical education in England, Wales and Scotland. With its general aim of promoting the advancement of British geography, the functions of the Council are:

1. To provide a forum for discussion and exchange between the organisations concerned with geography and geographical education at all levels (e.g. recent matters for discussion have included research funding; A/AS level Cores; supply of geography teachers).
2. To represent the interests of geography and geographers at national and international level (e.g. recent activities have included representations made about the Dearing National Curriculum review; also COBRIG's International Subcommittee has taken over the functions of the former British National Committee for Geography and so advises the Royal Society on International Geographical Union matters).
3. To identify issues which need investigation and/or concerted action on behalf of geography (e.g. recent issues include the National Curriculum; the character of geography in schools and higher education which was

Geography into the Twenty-first Century, Edited by E.M. Rawling and R.A. Daugherty,
© 1996 John Wiley & Sons Ltd.

explored at the Oxford seminar 1994; the quality of teaching and learning to be explored at the 1996 seminar).

The Council meets three times a year in the Council room of the Royal Geographical Society. It is financed by subscriptions from member bodies and a grant from the Royal Society to assist with international affairs. It has three honorary officers who conduct its day-to-day business. The present Honorary Chair is Professor Robert Bennett (London School of Economics), the Honorary Secretary is Neil Simmonds (Worksop College, Nottingham) and the Honorary Treasurer is Dr Tony Champion (University of Newcastle upon Tyne). Eleanor Rawling chaired the Council during 1993 and 1994, the period in which the seminar took place.

The present member bodies of Council are:

- The Royal Geographical Society with Institute of British Geographers
- The Geographical Association
- The Royal Scottish Geographical Society
- The Scottish Association of Geography Teachers
- The Scottish Council of Geography
- The University of Wales Council of Geography
- The Association of Geography Teachers of Wales
- The Conference of Heads of Geography in Higher Education Institutions
- Section E of the British Association for the Advancement of Science

Geographers in Northern Ireland are currently holding discussions with COBRIG about representation on the Council.

Appendix 2:
List of delegates present at the COBRIG seminar

New perspectives for geography in schools and in higher education, 2 and 3 July 1994, Oxford

Sara Bartley, Geography Teacher, Stratton Upper School, Biggleswade.

Jeff Battersby, Head of Geography, Parkside Community College, Cambridge, now Lecturer in Geographical Education, University of East Anglia [GA].

Linda Beskeen, Head of Geography, The Mary Redcliffe and Temple School, Bristol [RGS].

Dr Tony Binns, Senior Lecturer in Geography, University of Sussex [GA President].

Dr Michael Bradford, Senior Lecturer in Geography, University of Manchester [IBG].

Sue Burkill, Senior Lecturer in Geographical Education, College of St Mark and St John, Plymouth.

Graham Butt, Lecturer in Geographical Education, University of Birmingham.

Margaret Caistor, Geography Adviser, Islington and Consultant in Geographical Education.

Roger Carter, Senior Adviser (Geography) Staffordshire [GA, Chair of Education Standing Committee].

Graham Corney, Senior Lecturer in Geographical Education, University of Oxford.

Dr Alistair Cruickshank, Director, Royal Scottish Geographical Society, Glasgow.

Professor Richard Daugherty, Department of Education, University of Wales, Aberystwyth.

John Davidson, Head of Geography, Exeter School [GA].

Alex Dearing, Geography Teacher, Bishop Ullathorne RC School, Coventry.

Professor Derek Diamond, Department of Urban and Regional Planning, London School of Economics [President IBG 1994].

Alan Doherty, Head of Geography, Linlithgow Academy, Scotland [Editor *SAGT News*].

Martin Duddin, Head of Geography, Knox Academy, East Lothian [President SAGT 1994/95].

Dr William Edwards, Institute of Earth Sciences, Aberystwyth [UWCG].

Steve Frampton, Head of Geography, Peter Symonds College, Winchester.

Dr Rita Gardner, Director, Environmental Science Unit, Queen Mary and Westfield College, London.

Professor John Gold, Department of Geography, Oxford Brookes University [CHGHEI].

Professor Peter Haggett, Department of Geography, University of Bristol.

David Hall, School of Education, University of Bristol.

Dr Mick Healey, Senior Lecturer (Geography), Coventry University, now Professor at Cheltenham and Gloucester College of Higher Education [IBG].

Dr John Hemming, Director, Royal Geographical Society.

Leslie Hunter, HM Inspectorate, Scotland.

Elspeth Insch, Headteacher, King Edward VI School, Birmingham [RGS, Chair Education Committee].

Professor Peter Jackson, Department of Geography, University of Sheffield.

Professor Ron Johnston, Vice Chancellor, University of Essex now Professor of Geography, University of Bristol.

Mathew Judd, Geography Teacher, Haberdashers' Aske's School, Elstree.

Ashley Kent, Lecturer in Geographical Education, Institute of Education, London [GA].

Chris Kington, Publisher, Longman Logotron now Managing Director of Chris Kington Publishing.

Professor Eleonore Kofman, Department of International Studies, Nottingham Trent University [CHGHEI].

Jeremy Krause, Senior Adviser (Geography), Cheshire [GA].

David Lambert, Senior Lecturer in Geographical Education,

Institute of Education, London [GA].

Elizabeth Lewis, School of Education, University of Sunderland.

Christine McCullough, Economic and Social Science Research Council [IBG, Chair of Education Committee].

Averil Mander, School of Geography, University of Oxford.

Wendy Morgan, Consultant in Geographical Education and Editor, Primary Geographer [GA].

Michael Morrish, Head of Geography, Alleyns School, Dulwich, now Head of Geography, Haberdashers' Aske's School, Elstree [GA].

Professor Derek Mottershead, Department of Geography, Edge Hill College of HE [Honorary Secretary/COBRIG].

Professor Martin Parry, Environmental Studies Unit, University College, London.

Andrew Powell, Department of Geography, Kingston University [GA].

Professor Bruce Proudfoot, University of Glasgow [RSGS].

Graham Ranger, Adviser for Geography, Derbyshire [GA Honorary Secretary, Education].

Eleanor Rawling, Consultant in Geographical Education and Honorary Research Associate, University of Oxford, now Professional Officer (Geography) at School Curriculum and Assessment Authority [Chair, COBRIG].

Margaret Roberts, Department of Education, University of Sheffield.

Geoff Robinson, Director, Computers in Teaching Geography Initiative, University of Leicester.

Peter Smith, HMI (Geography), OFSTED.

Tony Smyth, Head of Geography, Chesham Park School, Chesham, High Wycombe [GA].

Christine Speak, Consultant in Geographical Education [GA].

Dr Iain Stevenson, Publisher, John Wiley and Sons, Chichester [IBG].

Professor Jim Taylor, Aberystwyth [UWCG].

Geraint Thomas, Head of Geography, Warden Park School, Haywards Heath, Sussex [GA].

Philip Thomas, Head of Geography, Ysgol-y-Gader, Dollgellau, Gwynedd, Wales [AGTW].

Dr Hilary Thomas, Lecturer, Gwent College of Higher Education [AGTW].

Dr Ramesh Tiwari, Department of Geography, University of Manitoba, Canada.

Dr Tim Unwin, Reader in Geography, Royal Holloway,

University of London [Honorary Secretary, IBG].

Rex Walford, Lecturer in Education (Geography), University of Cambridge [GA].

David Walker, Senior Lecturer in Geography, Loughborough University [IBG].

Stephen Watts, University of Sunderland [GA/Primary].

John Westaway, Professional Officer (Geography), School Curriculum and Assessment Authority, London.

Professor Brian Whalley, School of Geosciences, Queen's University, Belfast [Chair, BGRG].

Ann Whipp, Curriculum Council for Wales (CCW), Cardiff (now Curriculum and Assessment Authority for Wales (ACAC).

Abbreviations indicate representatives of:

AGTW	Association of Geography Teachers of Wales
BGRG	British Geomorphological Research Group
CHGHEI	Conference of Heads of Geography in Higher Education Institutions
GA	Geographical Association
IBG	Institute of British Geographers ⎫ now merged as
RGS	Royal Geographical Society ⎬ RGS/IBG from ⎭ 1 January 1995
RSGS	Royal Scottish Geographical Society
SAGT	Scottish Association of Geography Teachers
UWCG	University of Wales Council of Geography

Appendix 3:
Examination statistics 1990–95
(England, Wales and Northern Ireland)

Numbers of GCSE candidates, 1990–95 (England, Wales and Northern Ireland)

Subject	1990	% Change	1991	% Change	1992	% Change	1993	% Change	1994	% Change	1995
Science (Int.)*	216 091	141.87	522 663	39.59	729 560	5.02	766 217	14.69	878 774	13.53	997 675
Mathematics	663 734	−4.83	631 704	0.79	636 716	−24.06	483 506	32.49	640 608	5.91	678 445
English	679 411	3.28	657 108	0.79	655 930	−1.67	465 004	−6.17	605 182	7.24	648 987
Engligh Lit.	428 810	3.94	455 717	1.69	453 234	−0.23	452 211	−5.08	429 226	10.63	474 846
French	281 576	8.55	305 660	5.85	323 535	−1.15	319 821	2.64	328 266	9.89	360 743
Geography	274 237	−1.70	269 582	0.63	271 274	−3.35	262 190	0.93	264 636	5.67	279 649
History	229 507	−5.44	217 020	0.58	218 279	−2.58	223 908	4.62	234 252	5.84	247 929
Art and Design	221 612	−9.03	201 610	9.51	220 792	−3.11	213 920	−0.66	212 504	0.04	212 583
CDT	172 940	5.72	182 836	−37.59	114 109	34.66	153 661	26.59	194 520	−53.77	89 933
German	84 306	−8.27	91 277	11.08	101 388	6.92	108 401	9.75	118 972	8.75	129 386
Home Economics	150 452	−13.13	130 694	−9.57	118 182	−6.97	109 948	−3.72	105 859	−48.29	54 736
Biology	229 775	−30.11	160 595	−35.72	103 233	−11.69	91 169	−17.53	75 183	−21.64	58 917
Physics	190 004	−30.48	132 095	−40.92	78 037	−16.35	65 276	−18.04	53 503	−18.06	43 839
Chemistry	173 723	−29.21	122 986	−41.35	2 133	−13.63	62 302	−16.45	52 056	−15.63	43 921
Total	3 996 178	1.89	4 071 547	0.618	4 096 402	−3.39	3 957 534	5.96	4 193 541	3.054	4 321 589

Source: School Curriculum and Assessment Authority, 1996

Numbers of A level candidates, 1990–95 (England, Wales and Northern Ireland)

Subject	1990	% Change	1991	% Change	1992	% Change	1993	% Change	1994	% Change	1995
English	74 172	6.75	79 187	9.59	86 779	2.83	89 238	-1.15	88 214	-1.98	86 467
Social Science	56 176	9.74	61 649	14.18	70 393	8.11	76 103	4.01	79 154	-1.36	78 074
Mathematics	79 784	-5.99	74 972	-3.45	72 384	-8.35	66 340	-2.14	64 919	-0.48	64 605
General Studies	51 399	1.33	52 085	3.15	53 724	2.11	54 858	-2.24	53 630	7.25	57 520
Biology	46 465	0.31	46 507	4.58	48 742	-2.04	47 748	6.50	50 851	2.76	52 255
Geography	41 671	4.60	43 586	4.74	45 653	1.81	46 479	-0.28	46 347	-6.24	43 454
History	43 808	0.52	44 034	6.05	46 698	-0.91	46 274	-3.34	44 730	-2.09	43 796
Chemistry	46 197	-3.80	44 440	-3.92	42 697	-4.03	40 975	0.62	41 231	2.58	42 293
Physics	45 334	-4.23	43 416	-4.87	41 301	-7.59	38 168	-5.30	36 147	-3.72	34 802
Art and Design	31 735	-1.70	31 195	9.42	34 133	3.45	35 312	-1.59	34 749	-2.40	33 915
Economics	45 330	-4.79	43 160	-6.81	40 222	-9.43	36 428	-14.60	31 109	-14.49	26 500
French	27 245	13.03	30 794	1.52	31 261	-4.40	29 886	-3.16	28 942	-4.75	27 563
Total	589 290	0.99	595 325	3.17	613 987	-1.01	607 809	-1.28	600 023	-1.45	591 344

Source: School Curriculum and Assessment Authority, 1996

Author index

NB. This index contains names of personal authors and people treated as subjects. Institutional authors are included in the subject index.

Author index compiled by Margaret Binns

Subject index

Subject index compiled by Margaret Binns